PRAISE FOR AMIR D. ACZEL and *THE MYSTERY OF THE ALEPH*

"*THE MYSTERY OF THE ALEPH* is a kind of guide to the terrain shared by theologians and mathematicians on the topic of infinity. . . . Along the way, Mr. Aczel reminds us of some of the truly fabulous paradoxes and puzzles that thinkers have wrestled with."

—*The Wall Street Journal*

"Aczel's biographical armatures, his clean prose, and his asides about Jewish mysticism keep his book reader friendly. It's a good introduction to an amazing and sometimes baffling set of problems, suited to readers interested in math—even, or especially, if they lack training."

—*Publishers Weekly*

"Theoretical mathematicians will revel in this well-written, witty book and even non-mathematicians will be carried along by a narrative with the pace of a thriller."

—*The New Scientist*

"What makes this book fascinating is not the math but the concepts drawn from the numbers and the stories of those who created them. It reveals a world that few enter, but for those who do, who understand, and who survive what is going on, it must be rather compelling."

—*The Seattle Times*

"In assessing Cantor's achievement as the first to probe infinity with mechanical rigor, Aczel demonstrates the same gift for interpreting complex concepts that he previously demonstrated in *God's Equation*, about Einstein's pioneering work in cosmology. And as in his book on Einstein, Aczel penetrates to the human mind behind the formulas, detailing the personal frustrations and professional conflicts that drove Cantor into mental collapse. Aczel also uncovers the uncanny ways in which Cantor's life foreshadowed that of his more famous successor, Gödel, who was attracted to the same problems and doomed to the same descent into madness."

—*Booklist* (starred review)

"[Aczel's writing is] a beautiful marriage of mathematics and prose."

—*Astronomy* magazine

"Aczel casts an intriguing light on Cantor's obsession."

—*The Boston Globe*

"Mr. Aczel is very good at portraying the essences of the thoughts and lives of that quirky class of geniuses known as mathematicians."

—*The New York Times*

"Aczel's language is blessedly clear."

—*The Boston Herald*

"An affecting life-and-work portrait. . . . Aczel's refined pop-maths antennae enable him to explain such bizarreries with pleasing allusions. . . . Concise and mind-stretching."

—*The Guardian* (UK)

"Sometimes technical but always accessible. . . . Aczel's book makes for a fine and fun exercise in brain stretching while providing a learned survey of the regions where science and religion meet."

—Amazon.com

Georg Cantor.

Credit: Martin-Luther-Universität, Halle-Wittenberg. UA Halle: Rep. 40 C11

The Mystery of the Aleph

Mathematics, the Kabbalah, and the Search for Infinity

Amir D. Aczel

WASHINGTON SQUARE PRESS
PUBLISHED BY POCKET BOOKS
New York London Toronto Sydney Singapore

A Washington Square Press publication of
POCKET BOOKS, a division of Simon & Schuster, Inc.
1230 Avenue of the Americas, New York, NY 10020

Published by arrangement with Four Walls Eight Windows

For information address Four Walls Eight Windows,
39 West 14th Street, New York, NY 10011

ISBN: 0-7434-2299-6

First Washington Square Press trade paperback printing
September 2001

10 9 8 7 6 5

For information regarding special discounts for bulk purchases,
please contact Simon & Schuster Special Sales at 1-800-456-6798 or
business@simonandschuster.com

Illustrations, unless otherwise noted, by Charles Nix

Cover design by John Fontana
Cover photos © Corbis, © Digital Art / Corbis

Printed in the U.S.A.

Contents

To Miriam, who at age 6 understood something about the difference between $\aleph_0$ and the power of the continuum.

$\aleph_0$

Halle

On January 6, 1918, an emaciated and weary man died of heart failure at the Halle Nervenklinik, a university mental clinic in the German industrial city of Halle. His body was quietly transported across town for burial in a small cemetery. Only a few people attended the Lutheran ceremony, including the man's widow and five surviving children.

The cemetery no longer exists; it has since been razed to make ground for private homes. But someone saved the headstone, and years later it was relocated, without the body, to another small burial ground in Halle, where it can still be seen. The chiseled inscription reads:

Dr. Georg Cantor
Professor d. Mathematik
3.3.1845–6.1.1918

By the time of his death, Georg Cantor had been hospitalized at the Halle Nervenklinik for seven months. But this stay at the clinic was not his first. Georg Cantor had been admitted, released, and readmitted to the clinic many times. And his mental problems had begun years before the clinic was built in 1891.

Georg Cantor received his doctorate in mathematics in 1869 from the University of Berlin, where he had studied under some of the world's greatest mathematicians and absorbed many important ideas in mathematics. He was eager to put his knowledge to use developing new theories in the field of mathematical analysis. The twenty-four-year-old was excited at the prospect of obtaining his first teaching position at a German university, hoping it would allow him time to pursue his research. But upon graduation, Cantor's only offer was from Friedrich's University in Halle, some seventy miles southwest of Berlin.

Halle is an old city with charming medieval cobblestoned streets. It was founded in the middle of the tenth century as a center of salt production on the Saale River. The city survived the bombings of the world war, and many ancient buildings still stand in the historic city center, where people stroll to shops and cafés unimpeded by motorized traffic. Halle is called the City of Five Towers. The four spires of the medieval Marktkirche loom over the lower buildings in the center of town, and nearby stands the fifth, the Red Tower of 1418, a monument to the struggle of the townspeople for independence from an oppressive aristocracy.

In 1685, the composer Georg Friedrich Handel was born in Halle in a house whose oldest standing walls date back to the twelfth century. Handel lived in this house for 18 years. The house, now a museum dedicated to the composer's life, can still be visited. Halle has always been a city of concerts, opera, and music of the people.

Credit: Fritz Moeller. Courtesy of Hanns H. Grote.

Marketplace of Halle an der Saale about 1900.

By all rights, Halle should have held some attraction for Cantor, as his family members on both sides were gifted musicians. Some of them had achieved renown in their native Russia. But Cantor was not interested in the charms of Halle. His was a family of immigrants—from the Iberian Peninsula via Denmark and Russia—and young Cantor was pushed to excel. His father, in particular, sent Georg letters throughout the years urging him to do well at school and to live up to the great expectations of his family.

Halle is situated halfway between two great university cities: Berlin to the northeast and Göttingen to the west. During the late nineteenth century, the University of Berlin was the world's best in mathematics, and Berlin was one of the most vibrant and exciting cities in all Europe. Göttingen was the other academic magnet. Like Halle, Göttingen is an old medieval city. Many houses in the town center bear plaques with the names of famous former residents, from Heine the poet to Bunsen the chemist to Olbers the astronomer, and many others, most notable among them Carl Friedrich Gauss (1777-1855), arguably the greatest mathematician of the time. Cantor felt the pull of both Berlin and Göttingen.

But Cantor stayed in Halle, waiting for the invitation that never came. Over the years, whenever a mathematics opening became available at Berlin or Göttingen he pinned his hopes on it, and when he wasn't offered the position, he would go into a fit of rage. He had an intense, demanding personality, and an explosive nature. These attributes made him enemies and lost him friends throughout his life. In contrast with his behavior with other mathematicians, Cantor exhibited tenderness in his relationships with his family members. While he always dominated conversations with colleagues, at home Cantor took a more relaxed role, letting his wife and children initiate and lead conversations at the dinner table. He ended every meal by asking his wife: "Have you been pleased with me today, and do you love me?"

Cantor started as a *Privatdozent*, the entry-level academic job at German universities of the time. Within a few years of hard work, he was promoted to Associate Professor, and shortly afterwards a Professor of Mathematics. Cantor became involved in intensive research in mathematics, but in the midst of his most productive period, something strange happened, which put a temporary end to his work. In the summer of 1884, Georg Cantor was struck by deep depression. From May through June of that year he was immobilized—unable to work or do much of anything. His condition distressed his wife and children and perplexed his colleagues, who saw in him a mathematician aspiring to great heights. However, without any professional help or medication, Cantor recovered from his illness and returned to normal life. Afterwards, he wrote a letter to a close friend, the Swedish mathematician Gösta Mittag-Leffler (1846-1927), describing his illness and mentioning that just before the mental breakdown he was working on the "continuum problem."

The following year, 1885, Cantor built an opulent house for his family on Handelstrasse, a street named after Halle's great composer. The house is still owned by Cantor's grandson. It is a two-story building, with high ceilings and tall windows. Georg Cantor's father, a merchant and stockbroker, had died a few years earlier, leaving his heirs half a million marks. Some of the inheritance money went into building the

Cantor's house.

Credit: Amir D. Aczel.

new house and buying furnishings so that the Cantor family could live in comfort. Today as then, Handelstrasse is a quiet, tree-lined street on which there are many expensive homes, and the house is ten minutes' walk from the university and from cafés, restaurants, and cultural institutions. But Cantor did not stay home with his family long enough to enjoy the new house. Shortly he fell ill again. This time, too, just before his mental breakdown, Cantor had been working on the continuum problem.

The University of Halle had an excellent department of psychiatry. Cantor could get the best treatment available at the time—and it was free, since he was a university professor. His university and the Ministry of Culture in Berlin, which authorized all such decisions, were generous in granting Cantor repeated leaves of absence from his teaching duties. But his hospitalizations became more frequent as the years went by. In the Prussian State Archives in Berlin there is a letter on budgets sent by the Culture Ministry to the Finance Ministry, dated August 29, 1902. In this letter, the Minister of Culture requests, among other things, an appropriation of 6,660 German marks to support a substitute appointment of Professor of Mathematics at the University of Halle in case Professor Dr. Cantor should be too ill to resume his duties. But Cantor recovered yet again and returned to teaching.

Within a year he was ill again, and was readmitted to the Nervenklinik on September 17, 1904, remaining there until March 1, 1905. Then in the fall of that year, Cantor was back in the clinic.

The Halle Nervenklinik is a complex of eleven buildings

constructed of attractive yellow-glazed bricks situated within a large fenced compound. The quality of construction was so high that the facility looks today almost exactly as it did when it was built over a century ago. The main building, with its pointed tower, resembles a military headquarters rather than a mental clinic. Inside, the rooms are spacious with large windows and private baths. This was not a place where people were restrained with straitjackets. It was—and still is—a clinic for short stays of several months by wealthy individuals whose families could afford room and board and treatment. Georg Cantor, a professor at the university, was given a single room with good view and had the freedom to pursue his research. His treatment consisted mainly of periods of soaking in a hot bath.

And although he did die while hospitalized at the clinic, there was certainly no justification for the statement Bertrand Russell later made about Cantor (referring to a letter Cantor had written) that those who will read his letter will not be surprised to hear that he died in an insane asylum.

We don't know the precise nature of Cantor's illness. Some of his reported symptoms resemble those associated with bipolar disorder, or manic depression. But the causes of this mental illness are now generally attributed to genetic factors, and in Cantor's ancestry there are no known cases of the disease.

One fact is known about Georg Cantor's illness. His attacks of depression were all associated with periods in which he was thinking about what is now known as "Cantor's continuum hypothesis." He was contemplating a single mathematical expression, an equation using the Hebrew letter aleph:

$$2^{\aleph_0} = \aleph_1$$

This equation is a statement about the nature of infinity. A century and a third after Cantor first wrote it down, the equation—along with its properties and implications—remains the most enduring mystery in mathematics.

$\aleph_1$

Ancient Roots

Sometime between the fifth and sixth centuries B.C., the Greeks discovered infinity. The concept was so overwhelming, so bizarre, so contrary to every human intuition, that it confounded the ancient philosophers and mathematicians who discovered it, causing pain, insanity, and at least one murder. The consequences of the discovery would have profound effects on the worlds of science, mathematics, philosophy, and religion two-and-a-half millennia later.

We have evidence that the Greeks came upon the idea of infinity because of haunting paradoxes attributed to the philosopher Zeno of Elea (495–435 B.C.). The most well-known of these paradoxes is one in which Zeno described a race between Achilles, the fastest runner of antiquity, and a tortoise. Because he is much slower, the tortoise is given a head start. Zeno reasoned that by the time Achilles reaches the point at which the tortoise began the race, the tortoise will have advanced some distance. Then by the time Achilles travels that new distance to the tortoise, the tortoise will have advanced farther yet. And the argument continues in this way ad infinitum. Therefore, concluded Zeno, the fast Achilles can never beat the slow tortoise. Zeno inferred from his paradox that motion is impossible under the assumption that space and time can be subdivided infinitely many times.

Another of Zeno's paradoxes, the dichotomy, says that you can never leave the room in which you are right now. First you walk half the distance to the door, then half the remaining distance, then half of what still remains from where you are to the door, and so on. Even with infinitely many steps—each half the size of the previous one—you can never get past the door! Behind this paradox lies an important concept: even infinitely many steps can sometimes lead to a *finite* total distance. If each step you take measures half the size of the previous one, then even if you should take infinitely many steps, the total distance traveled measures twice your first distance:

$$1+1/2 + 1/4 + 1/8 +1/16 +1/32 + 1/64+ \ldots\ldots = 2$$

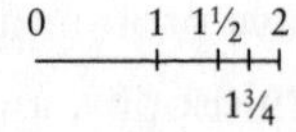

Zeno used this paradox to argue that under the assumption of infinite divisibility of space and time, motion can never even start.

These paradoxes are the first examples in history of the use of the concept of infinity. The surprising outcome that an infinite number of steps could still have a finite sum is called "convergence."

One could try to resolve the paradoxes by discarding the notion that Achilles, or the person trying to leave a room, must take smaller and smaller steps. Still, doubts remain, for if Achilles must take smaller and smaller steps, he can never win. These paradoxes point to disturbing properties of infinity and to the pitfalls that await us when we try to understand the meaning of infinite processes or phenomena. But the roots of infinity lie in the work done a century before Zeno by one of the most important mathematicians of antiquity, Pythagoras (c. 569-500 B.C.).

Pythagoras was born on the island of Samos, off the Anatolian coast. In his youth he traveled extensively throughout the ancient world. According to tradition, he visited Babylon and made a number of trips to Egypt, where he met the priests—keepers of Egypt's historical records dating from the dawn of civilization—and discussed with them Egyptian studies of number. Upon his return, he moved to Crotona, in the Italian boot, and established a school of philosophy dedicated to the study of numbers. Here he and his followers derived the famous Pythagorean theorem.

Before Pythagoras, mathematicians did not understand that results, now called theorems, had to be proved. Pythagoras and his school, as well as other mathematicians of ancient Greece, introduced us to the world of rigorous mathematics, an edifice built level upon level from first principles using axioms and logic. Before Pythagoras, geometry had been a collection of rules derived by empirical measurement. Pythagoras discovered that a complete system of mathematics could be constructed, where geometric elements corresponded with numbers, and where integers and their ratios were all that was necessary to establish an entire system of logic and truth. But something shattered the elegant mathematical world built by Pythagoras and his followers. It was the discovery of irrational numbers.

The Pythagorean school at Crotona followed a strict code of conduct. The members believed in metempsychosis, the transmigration of souls. Therefore, animals could not be slaughtered for they might shelter the souls of deceased friends. The Pythagoreans were vegetarian and observed additional dietary restrictions.

The Pythagoreans pursued studies of mathematics and philosophy as the basis for a moral life. Pythagoras is believed to have coined the words *philosophy* (love of wisdom) and *mathematics* ("that which is learned"). Pythagoras gave two kinds of lectures: one restricted to members of his society, and the other designed for the wider community. The disturbing finding of the existence of irrational numbers was given in the first kind of lecture, and the members were sworn to complete secrecy.

The Pythagoreans had a symbol—a five-pointed star enclosed in a pentagon, inside of which was another pentagon, inside it another five-pointed star, and so on to infinity. In this figure, each diagonal is divided by the intersecting line into two unequal parts. The ratio of the larger section to the smaller one is the golden section, the mysterious ratio that appears in nature and in art. The golden section is the infinite limit of the ratio of two consecutive members of the Fibonacci series of the Middle Ages: 1, 1, 2, 3, 5, 8, 13, 21, 34, 55, 89, 144, 233, . . . where each number is the sum of its two predecessors. The ratio of each two successive numbers approaches the golden section: 1.618. . . . This number is irrational. It has an infinite, nonrepeating decimal part. Irrational numbers would play a crucial role in the discovery of orders of infinity two and a half millennia after Pythagoras.

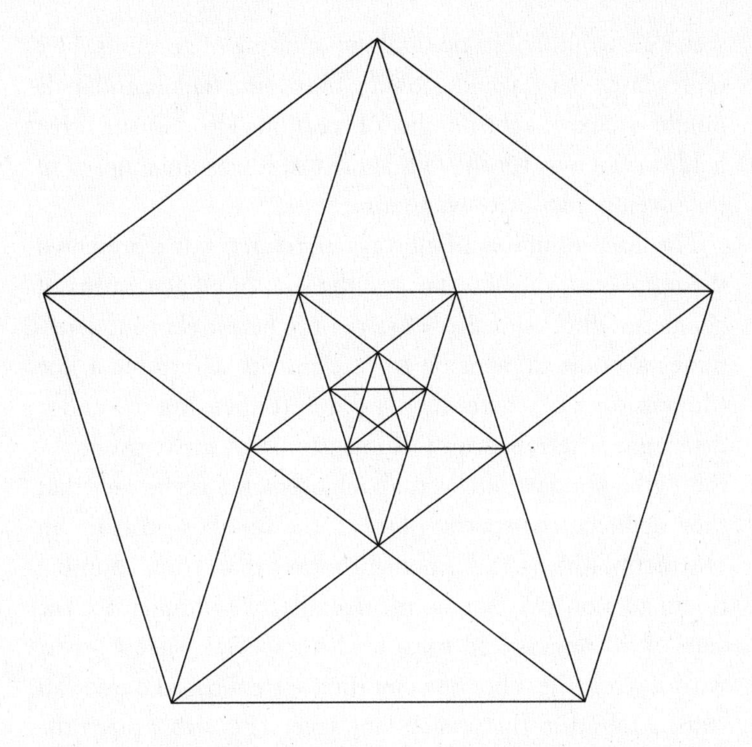

Number mysticism did not originate with the Pythagoreans. But the Pythagoreans carried number-worship to a high level, both mathematically and religiously. The Pythagoreans considered one as the generator of all numbers. This assumption makes it clear that they had some understanding of the idea of infinity, since given any number—no matter how large—they could generate a larger number by simply adding one to it. Two was the first even number, and represented opinion. The Pythagoreans considered even numbers female, and odd numbers male. Three was the first true odd number, representing harmony. Four, the first square, was seen as a symbol of justice and the squaring of accounts. Five represented marriage: the joining of the first female and male numbers. Six was the number of creation. The number seven held special awe for the Pythagoreans: it was the number of the seven planets, or "wandering stars."

The holiest number of all was ten, *tetractys*. It represented the number of the universe and the sum of all generators of geometric dimensions: 10=1+2+3+4, where 1 element determines a point (dimension 0), 2 elements determine a line (dimension 1), 3 determine a plane (dimension 2), and 4 determine a tetrahedron (3 dimensions). A great tribute to the Pythagoreans' intellectual achievements is the fact that they deduced the special place of the number 10 from an abstract mathematical argument rather than from counting the fingers on two hands. Incidentally, the number 20, the sum of all fingers and toes, held no special place in their world, while the relics of a counting system based on 20 can still be found in the French language. This strengthens the argument that the Pythagoreans made inferences based on abstract mathematical reasoning rather than common anatomical features.

Ten is a *triangular* number. Here again we see the strong connection the Pythagoreans saw between geometry and arithmetic. Triangular numbers are numbers whose elements, when drawn, form triangles. Smaller triangular numbers are three and six. The next triangular number after ten is fifteen.

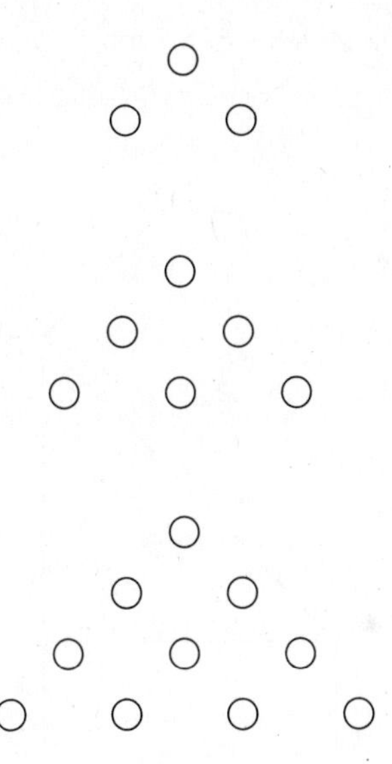

A later Pythagorean, Philolaos (4th c. B.C.) wrote about the veneration of the triangular numbers, especially the tetractys. Philolaos described the holy tetractys as all-powerful, all-producing, the beginning and the guide to divine and terrestrial life.[1] Much of what we know about the Pythagoreans comes to us from the writings of Philolaos and other scholars who lived after Pythagoras.

The Pythagoreans discovered that there are numbers that cannot be written as the ratio of two whole numbers. Numbers that cannot be written as the ratio of two integers are called *irrational* numbers. The Pythagoreans deduced the existence of irrational numbers from their famous theorem, which says that the square of the hypotenuse of a right triangle is equal to the sum of the squares of the other two sides, $a^2 + b^2 = c^2$. This is demonstrated in the figure below.

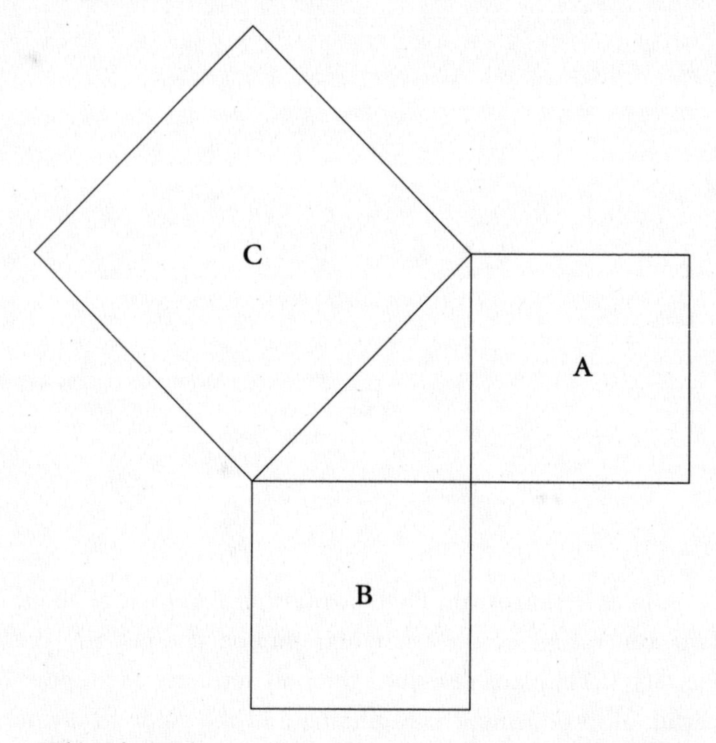

When the Pythagorean theorem is applied to a triangle with two sides equal to one, the result is that the hypotenuse is given by the equation $c^2 = 1^2 + 1^2 = 2$, so that $c = \sqrt{2}$. The Pythagoreans realized that this new number could not possibly be written as the ratio of two integers, or whole numbers.[2] Rational numbers, which are of the form *a*/*b* where both *a* and *b* are integers, have decimals that either become zero eventually, or have a pattern that repeats itself indefinitely. For example, 1/2=0.50000 . . . ; 2/3=0.6666666 . . .; 6/11=0.54545454. . . . Irrational numbers, on the other hand, have decimals that do not repeat the same pattern. Thus to write them exactly one would need to write infinitely many decimals.[3]

The irrationals were a devastating discovery for Pythagoras and his followers because numbers had become the Pythagoreans' religion. *God is number* was the cult's motto. And by *number* they meant whole numbers and their ratios. The existence of the square root of two, a number that could not possibly be expressed as the ratio of two of God's creations, thus jeopardized the cult's entire belief system. By the time this shattering discovery was made, the Pythagoreans had become a well-established society dedicated to the study of the power and mystery of numbers.

Hippasus, one of the members of the Pythagorean order, is believed to have committed the ultimate crime by divulging to the outside world the secret of the existence of irrational numbers. A number of legends record the aftermath of the affair. Some claim that Hippasus was expelled from the society. Others tell how he died. One story says that Pythagoras himself strangled or drowned the traitor, while another describes how the Pythagoreans dug a grave for Hippasus while he was still alive and then mysteriously caused him to die. Yet another legend has it that Hippasus was set afloat on a boat that was then sunk by members of the society.

In a sense, the Pythagoreans' idea of the divinity of the integers died with Hippasus, to be replaced by the richer concept of the continuum. For it was after the world learned the secret of the irrational numbers that Greek geometry was born. Geometry deals with lines and planes and angles, all of which are continuous. The irrational numbers are the natural inhabitants of the world of the continuum—although rational numbers live in that realm as well—since they constitute the majority of numbers in the continuum. A rational number can be stated in a finite number of terms, while an irrational number, such as π (the ratio of the circumference of a circle to its diameter), is intrinsically infinite in its representation: to identify it completely, one would have to specify an infinite number of digits. (With irrational numbers there is no possibility of saying: "repeat the decimals 17342 forever," since irrational numbers have no patterns that repeat forever.)

Pythagoras died in Metapontum in southern Italy around 500 B.C., but his ideas were perpetuated by many of his disciples who dispersed throughout the ancient world. The center at Crotona was abandoned after a rival mystical group called the Sybaris mounted a surprise attack on the Pythagoreans and murdered many of them. Among those who fled, carrying Pythagoras's flame, was a group that settled in Tarantum, farther inland in the Italian boot than Crotona. Here Philolaos was trained in the Pythagorean number mysticism in the following century. Philolaos's writings about the work of Pythagoras and his disciples brought this important body of work to the attention of Plato in Athens. While not himself a mathematician, the great philosopher was committed to the Pythagorean veneration of number. Plato's enthusiasm for the mathematics of Pythagoras made Athens the world's center for mathematics in the fourth century B.C. Plato became known as the "maker of mathematicians," and his academy had at least four members considered among the most prominent mathematicians in the ancient world. The most important one for our story was Eudoxus (408–355 B.C.).

Plato and his students understood the power of the continuum. In keeping with number worship—now brought to a new level—Plato wrote above the gates of his academy: "Let no one ignorant of geometry enter here." Plato's dialogues show that the discovery of the incommensurable magnitudes—the irrational numbers such as the square roots of two or five—stunned the Greek mathematical community and upset the religious basis of the Pythagoreans' number worship. If integers and their ratios could not describe the relationship of the diagonal of a square to one of its sides, what could one say about the perfection the sect had attributed to whole numbers?

The Pythagoreans represented magnitudes by pebbles or *calculi*. The words "calculus" and "calculation" come from the calculi of the Pythagoreans. Through the work of Plato's mathematicians and Euclid of Alexandria (c. 330–275 B.C.), author of the famous book *The Elements*, magnitudes became associated with line segments, as arithmetized geometry took the place of the calculi. The dichotomy between numbers and continuous magnitudes required a new approach to mathematics—as well as to philosophy and religion. In keeping with this new way of seeing things, Euclid's *Elements* discussed the solution of a quadratic equation, for example, not algebraically but as an application of areas of rectangles. Numbers still reigned in Plato's academy, but now they were viewed in the wider context of geometry.

In the *Republic*, Plato says "Arithmetic has a very great and elevating effect, compelling the mind to reason about abstract number." *Timaeus*, a book in which Plato writes about Atlantis, is named after a member of the Pythagorean order. Plato also refers to a number he calls "the lord of better and worse births," a number that through the centuries has become the subject of much speculation. But Plato's greatest contribution to the history of mathematics lies in having had disciples who advanced the understanding of infinity.

Zeno's idea of infinity was taken up by two of the greatest mathematicians of antiquity: Eudoxus of Cnidus (408–355 B.C.) and Archimedes of Syracuse (287–212 B.C.). These two Greek mathematicians made use of infinitesimal quantities—numbers that are infinitely small—in trying to find areas and volumes. They used the idea of dividing the area of a figure into small rectangles, then computing the areas of the rectangles and adding these up to an approximation of the unknown desired total area.

Eudoxus was born to a poor family, but had great ambition. As a young man, he moved to Athens to attend Plato's Academy. Too poor to afford life in the big city, he found lodgings in the port town of Piraeus, where the cost of living was low, and commuted daily to the academy in Athens. Eudoxus became Plato's star student and traveled with him to Egypt. Later in his life, Eudoxus became a physician and legislator and even contributed to the field of astronomy.

In mathematics, Eudoxus used the idea of a limit process. He found areas and volumes of curved surfaces by dividing the area or volume in question into a large number of rectangles or three-dimensional objects and then calculating their sum. Curvature is not easily understood, and to compute it, we need to view a curved surface as the sum of a large number of flat surfaces. Book V of Euclid's *Elements* describes this, Eudoxus's greatest achievement: the method of exhaustion, devised to compute areas and volumes. Eudoxus demonstrated that we do not have to assume the actual exis-

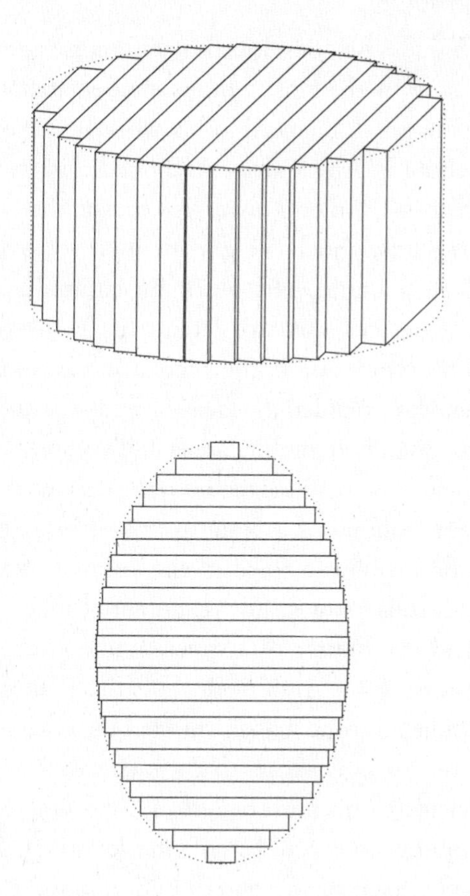

tence of *infinitely* many, infinitely small quantities used in such a computation of the total area or volume of a curved surface. All we have to assume is that there exist quantities "as small as we wish" by the continued division of any given total magnitude: a brilliant introduction of the concept of a *potential infinity.* Potential infinity enabled mathematicians to develop the concept of a limit, developed in the nineteenth century to establish the theory of calculus on a firm foundation.

The techniques first developed by Eudoxus were expanded a century later by the most famous mathematician of antiquity: Archimedes. Influenced in his work by ideas of Euclid and his school in Alexandria, Archimedes is credited with many inventions. Among his discoveries is the famous law determining how much weight an item loses when it is immersed in a liquid. His work on catapults and other mechanical devices used to defend his beloved Syracuse enhanced his reputation in the ancient world. In mathematics, Archimedes extended the ideas of Eudoxus and made use of potential infinity in finding areas and volumes using infinitesimal quantities. By these methods, he derived the rule stating that the volume of a cone inscribed in a sphere with maximal base equals a third of the volume of the sphere. Archimedes thus showed how a potential infinity could be used to find the volume of a sphere and a cone, leading to actual results. After Archimedes' death at the hands of a Roman soldier, a stone mason chiseled the cone inscribed in a sphere on his gravestone to commemorate what Archimedes considered his most beautiful discovery.

Greek philosophers and mathematicians of the Golden Age, from Pythagoras to Zeno to Eudoxus and Archimedes, discovered much about the concept of infinity. Surprisingly, for the next two millennia, very little was learned about the mathematical properties of infinity. The concept of infinity, however, was reborn during medieval times in a new context: religion.

ℵ$_2$

Kabbalah

When the Israelites left Egypt in the second millennium B.C., they established the Jewish priesthood. The first to hold the position of head priest was Aaron, Moses's older brother. The priest wore around his neck a golden chain with a rectangular array of twelve squares of precious metals, each symbolizing one of the Twelve Tribes of Israel. This ceremonial plate, called the *Urim veTumim*, was believed to possess strong mystical powers. The Urim veTumim helped see the Israelites through their forty-year ordeal in the desert. The Israelites used the Urim veTumim in the ceremony in which the Ten Commandments were handed down at Mount Sinai, then carried it with them throughout their conquest of the Holy Land, finally to place it in the Temple in Jerusalem. With the priesthood and the Urim veTumim, Jewish mysticism was born.

A thousand years later, when the Jews returned from the Babylonian Exile, scribes wrote down secret interpretations of the hidden meanings in the Torah. These writings were highly allegorical and their study was entrusted to a select group of scholars. The writings were elaborated and expanded after the beginning of the Second Exile following the Roman destruction of the Temple in Jerusalem in A.D. 70.

After this traumatic event, the Jewish leadership dispersed in Judaea, and a number of sages settled in the town of Yavne, away from the city of Jerusalem in which Jews were now forbidden to reside. These first rabbis, replacing the priests of the temple, established an academy of learning. Among them was someone who was to become a great spiritual leader of the Jews: Rabbi Joseph ben Akiva (c. A.D. 50–132).

Rabbi Akiva wrote a collection of papers called *Maaseh Merkava*, or The Way of the Chariot. The rabbi's writings taught the believers a new way to spirituality. His method consisted of creating visual images of heavenly palaces, whose purpose was to induce meditation and through it closeness to the Divine.

Rabbi Akiva had apparently chanced upon a practice that was almost too intense for the human mind. The meditations the rabbi prescribed called for inducement of out-of-body experiences, altered mental states, and heights of ecstasy not previously known in Western culture. While the visions of the heavenly palaces on the way to the One were vivid and intense, Rabbi Akiva exhorted his students not to succumb to hallucinations or lose their grasp on reality. "When you enter the pure stones of marble [a stage of meditation]," he wrote, "do not say 'Water! Water!' for the Psalm tells us, 'He who speaks falsely will not be established before my eyes.'"

The rabbi used biblical passages and chants he composed himself as vehicles for achieving meditative states of mind.

One of these devices was an infinitely bright light the students visualized, symbolizing the *chaluk*, or robe, which covered God when he appeared to Moses on Mount Sinai. In their meditations, the students strove to achieve the intensity of Moses as he witnessed the robed figure of God.

According to legend, Rabbi Akiva and three of his colleagues entered the palaces of the meditation together. Their experience was so intense that the first, Rabbi Ben Azai, gazed at the infinite light and died, for his soul so longed for the source of light that he instantly shed his physical body and was no more. The second, Rabbi Ben Abuya, looked at the divine light and saw two gods instead of one. He became an apostate. The third, Rabbi Ben Zoma, glanced at the infinite light of God's robe and lost his mind, for he could not reconcile ordinary life with his vision. Only Rabbi Akiva survived the experience.

The work of Rabbi Akiva was studied by generations of Jewish scholars in the Diaspora over the following centuries. These studies took place in strict secrecy for a number of reasons. First, the intensity of the experiences was considered dangerous for the inexperienced. And second, the Jews were not masters of their land—whether in Palestine or the lands of the Diaspora—and their rulers might not have looked favorably on Jews dabbling in mysticism. The mystics therefore covered up their work and often distorted their writings to confuse the uninitiated. To insure the integrity of the tradition, it was passed orally from master to student.

In the tenth century, the Babylonian school of Hai Gaon (A.D. 939–1038) focused the meditations introduced by Rabbi Akiva and his followers on individual expansion of spiritual consciousness rather than on altered mental states. In Palestine and in Europe, the mystical meditations remained in the spirit of the *Maaseh Merkava*. The guiding principle of the meditations was that, through them, any Jew would be able to replicate the experience of being present at Mount Sinai when the Israelites received the Ten Commandments.

In the eleventh century in Spain, the mystic Solomon Ibn Gabirol gave a name to the Jewish system of secret mysticism and meditations. He called it *Kabbalah*: the "received" tradition of the Jews. These were the secret teachings, received from mouth to ear, a direct transmission of timeless spiritual wisdom. In Spain and elsewhere, the Kabbalists, as they were now called, organized themselves as a secret society dedicated to the study of the ancient wisdom of the Torah and the commentaries, looking for mysterious connections and hidden truths. Soon they turned their attention to number.

Each letter in the Hebrew alphabet was assigned a numerical value. The sages held that words that had the same total numerical sum of letters were connected in some way. This study of number and associated meaning was known as *gematria*. Permutations of letters in the Hebrew alphabet were used to study hidden meanings in the Torah (the five books of Moses) by early Kabbalists in Palestine. In the twelfth century, French Kabbalists added to this practice meditations based on the Tetragrammaton—the four-letter name of God, YHVH. These meditations included breathing exercises and bodily gestures in addition to the study of numerical values of the letters of the Tetragrammaton.

In 1280, the Spanish Kabbalist Moses de Léon wrote a master study on Kabbalah in which he put together all the important elements of meditation and mysticism known from ancient times to his own era. The book was called the *Zohar*, meaning radiance—or intense light, the light of the infinity of God. The Zohar was a result of de Léon's mystical experience, the outcome of his meditations on the Divine Name. It was written in the ancient, enigmatic language of Aramaic, the lingua franca of the Near East at the dawn of civilization. The Zohar's roots are steeped in the early tradition of the Kabbalah of Shimon Bar Yohai, a student of Rabbi Akiva. To this day, the Zohar is the most important book of the Kabbalah.

But the genesis of the Zohar is shrouded in mystery and has been the subject of an ongoing controversy throughout the centuries. Moses de Léon started by circulating parts of what he later compiled as the Zohar among his friends, claiming that these were ancient writings of Shimon Bar Yohai which he was simply transcribing. According to tradition, Bar Yohai spent twelve years in a cave in the Galilee writing his meditations, which were later hidden during the Roman occupation of Israel and eventually smuggled out and into Spain. In 1305, however, a refugee from the Mamluk siege of Acre, a city on the Mediterranean coast of the Holy Land, arrived in Spain. He claimed to know all ancient scripts extant in Palestine and concluded that de Léon's pamphlets could not have reproduced the ancient writings of Bar Yohai but were a contemporary work.

Other inquiries found evidence to support de Léon's claim that the writings were ancient. Within a few years, the pamphlets were put together in volumes of the Zohar. Between 1280 and 1286, Moses de Léon produced the entire voluminous work. Doubts about the manuscript's authenticity persisted through the centuries. But whether the Zohar is of ancient provenance or a thirteenth-century creation based on ancient ideas, the document is immensely important in Jewish mysticism. The Zohar forms the backbone of the mystical philosophy of Kabbalah.

With the advent of printing, the book became more accessible and its first printed editions were made in Mantua and Cremona in Italy in 1558 and 1560. The controversy about the mystical writings was reignited with the increased availability of the text. A number of Kabbalists believed that it was dangerous and forbidden to divulge the hidden secrets of the holy Torah by publishing the Zohar. Others opposed publication on the grounds that the manuscript was not an ancient text.

A century later, more controversy erupted with the rise of a man who became known as the False Messiah, Shabbetai Zevi. The Sabbatianist movement of the seventeenth century, headed by Shabbetai, drew much of its imagery, symbolism and doctrine from the Zohar. Shabbetai was arrested in Turkey in 1666 and given a choice of death or conversion to Islam. He chose the latter. The power of the imagery of the Zohar was so sweeping, however, that there was no stopping its spread, and the movement continued throughout the eighteenth century.

Also in thirteenth-century Spain, Abraham Abulafia opened the practice of Kabbalah to women and non-Jews, in what became one of the most controversial developments in the discipline. He came into conflict both with the Jewish religious authorities and the Inquisition. His new approach prepared the way for messianic cults that were to dominate Jewish mysticism for centuries.

In the 1500s, a number of great Kabbalists, fleeing the Spanish Inquisition and motivated by a desire to be close to the holy places of their people, moved to the hill town of Safed in the Galilee. To this day, Safed has the air of a holy place. The outskirts of the city are dotted with the tombs of famous medieval rabbis, and the cobblestoned walkways of the city still look as they did half a millennium ago. Religious scholars still walk the ancient streets dressed in the black garb they wore then, except that now most of them talk on their cellular phones. The city houses some of the most ancient standing synagogues. Here, under the leadership of the *Ari* (the Lion) Isaac Luria (1543–1620), a secret community of mystics—much like the Pythagoreans of two thousand years earlier—was established. The community's members, called the *Chaverim*, or friends, shared duties, prayers, meals, and meditations. The Ari introduced a new method of meditation, the *tikkun*, or repair, a form of deep concentration aimed at binding the world of form to the Absolute. Luria assigned each of his students a unification exercise suited to the student's individual characteristics. To heighten each student's meditative state, the Ari used incense, fragrant herbs, and spices. These aids were designed to allow the students a chance to approach the sublimeness of God.

The community flourished and produced some of the greatest names in Kabbalah, among them Moses Cordovero (1522–1570), who at the time was the leading Kabbalist of Safed. He instituted group meditations at the graves of sages and led scriptural discourses. Another important scholar was Joseph

Caro, whom a spirit ordered to go from Spain to Safed, where he arrived in 1536. Caro, Cordovero, and their colleagues instituted Hebrew as the language of the community, and taught the members to recite the oral tradition with its Kabbalistic interpretations. The Zohar itself arrived in Safed in 1570 with Luria, who traveled there from Egypt. In Safed, under the influence of so many keen scholars, the essential elements of Kabbalah as it is known and practiced today were established. What are these elements?

At the heart of the Kabbalah lie ten *Sefirot*, or "countings." Thus the number ten holds a special meaning for Jewish mystics, as it did for the Pythagoreans in the form of the holy tetractys. (There are also ten Jewish martyrs mentioned in penitential prayer—one of whom is Rabbi Akiva.)

The Sefirot, the hidden elements of Kabbalah, are arranged in a mysterious geometrical shape as ten spheres occupying multi-dimensional space. When drawn on paper, they appear as shown below, but are understood to extend into more dimensions.

Each of the four letters of the Hebrew name of God represents a world. The ten permutations of the letters form the Sefirot. The first is *Keter*, meaning crown. It represents will and humility and fifth-dimensional consciousness, and is associated with the colors white and black. The second is *Binah*, meaning understanding. It represents happiness and is associated with the color green. The third *Sefira* is *Chochma*, meaning wisdom. It also represents egolessness and is associated with the color blue. *Gevura* means might or heroism, and is associated with strength, awe, restraint and the color red. *Chesed* is mercy or kindness and is asso-

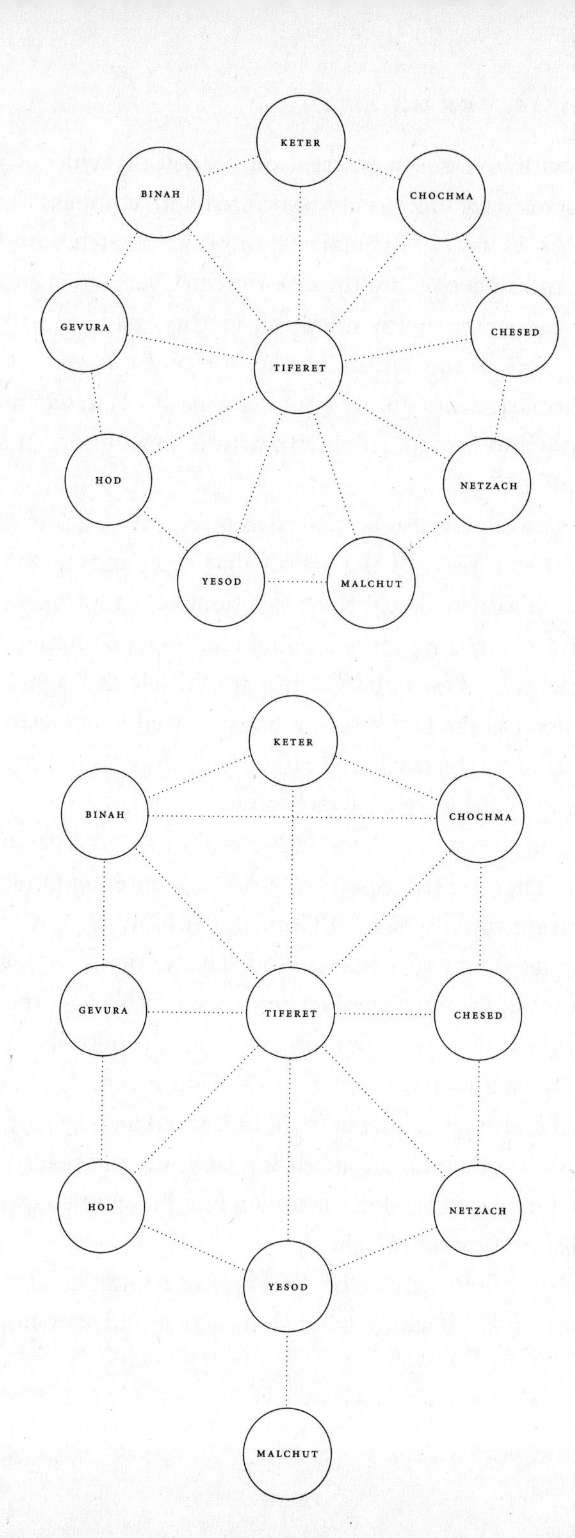
KETER
BINAH
CHOCHMA
GEVURA
CHESED
TIFERET
HOD
NETZACH
YESOD
MALCHUT
KETER
BINAH
CHOCHMA
GEVURA
TIFERET
CHESED
HOD
NETZACH
YESOD
MALCHUT

ciated with love and with creation. Its color is white. *Tiferet* is beauty or elegance and is associated with compassion and the color white. *Hod* is majesty and is associated with flexibility and sincerity and the color green. *Netzach* is eternity and is associated with splendor, victory, and security. Its color is red. *Yesod* means foundation and represents truth and formation. Its color is white. Finally, *Malchut* means kingdom and is associated with action, perception, and the color white.

The Sefirot may also be viewed as representing a primordial human form. Each of the Sefirot thus represents a different limb of this divine body. Keter is a human head; Chochma is a male face, and Binah is a female face—representing mind and intellect. Chesed and Gevura are the left and right arms, and Tiferet is the trunk of the body as well as its heart and spinal column. Netzach and Hod are the legs, Yesod the genitals, and Malchut represents both feet.

Each of the ten Sefirot stands for a set of qualities of the Divine. They are ten aspects of God and represent guidelines in aspiring to closeness to God. But behind the ten Sefirot stands the great entity that is God. That entity is so large, so supreme, so far beyond description, that it is given the only name the Kabbalists could possibly use to describe it: *Ein Sof*. The two words mean *Infinity*. God is infinite. The Ein Sof is the ultimate concept in all of Kabbalah. The first Kabbalist to use the name Ein Sof for God was the twelfth century rabbi Isaac the Blind. It took a blind man to conceive of the idea of an infinite light.

God as infinity cannot be described or comprehended. The Ein Sof is far beyond what a human mind can hope to glimpse. A fourteenth-century Kabbalist, whose identity is unknown, wrote: "Ein Sof is not hinted at in the Torah, Prophets, or Writings, neither in the words of our Rabbis, may their memory be blessed, but the Masters of the Service (the Kabbalists) have received a little hint of It."[4]

One form in which the infinite appears in the Zohar is the following:

> When the King conceived ordaining
> He engraved engravings in the luster on high.
> A blinding spark flashed
> Within the Concealed of the Concealed
> From the mystery of the Infinite,
> A cluster of vapor in formlessness,
> Set in a ring,
> Not white, not black, not red, not green,
> No color at all . . .[5]

Since God is Infinity and cannot be comprehended, the Sefirot are the finite aspects that the Kabbalists have gleaned from the immensity of the Ein Sof. The attributes of the Sefirot can be studied and meditated upon, and prayed with.

The Sefirot embody God's creation, from the lowliest level of a mineral all the way to the wonder of the great Ein Sof itself. Moses de Léon wrote in the Zohar: "God is unified oneness. Down to the last link, everything is tied together with everything, so divine essence is below as well as above, in heaven and on Earth." The Zohar speaks in terms of lights and roots and crowns and garments as images of the individual Sefirot. The reader must interpret the rich mystical images and try to identify each Sefirah.

With the appearance of the Sefirot and the Ein Sof early in the thirteenth century, the Kabbalists were charged with polytheism. How could God be both infinite and ten Sefirot? The Kabbalists responded that God is infinite, the Ein Sof, but the Sefirot are parts of that Ein Sof, forming a unity "like a flame joined to a coal." While the Sefirot appear to have a multiple existence, all of them are one and form part of the Infinite.

Thus the Kabbalists seem to have had a firm grasp of the concept of infinity, much as the Greek philosophers and mathematicians did. They understood that infinity could contain finite parts, but that the whole, infinity itself, was immeasurably greater than its parts. In their response to the charge of polytheism, the Kabbalists have used an important concept that also appears in the modern theory of sets. It is called the problem of *the one and the many*.[6] The problem arises when we ask the question: When can many objects be considered as one, that is, as a *set* containing all the individual objects? This is a difficult problem since it leads to paradoxes such as the famous Russell's paradox, discussed later.

One of the ideas of the Kabbalah is nothingness. The Kabbalists encountered the same problem that stumps anyone who tries to imagine pure nothingness. People always want to visualize *something*—a box, a container, or empty space—containing nothing. In Kabbalah that container is called a vessel or a garment. In modern set theory we consider the *empty set*, a set containing nothing. The arguments in the Kabbalah concerning the vessel lead to the same conclusions and paradoxes that plague twenty-first-century set theory. These considerations affect our understanding of infinity.

The Kabbalah uses the concept of Ein Sof in more than one context. There is a discrete infinity of elements (of which the ten Sefirot are part), including the integers. Then, there are passages in Kabbalah writings in which scholars discuss lines extending endlessly and curving toward a point at infinity. This is the *continuous* infinity of geometry and Plato and his disciples, rather than that of the early Pythagoreans before the discovery of the irrational numbers. Thus the Kabbalists were apparently aware of the fact that infinity exists both as an endless collection of discrete items and as a continuum. God was viewed as both of these infinities, as well as infinities so complex that they could not be conceived by the human mind. But the mathematical understanding and development of the different kinds of infinity would have to wait.

In its continuous form, Ein Sof appears as an infinite ray of light of infinite intensity. The light fills space and curves around it toward infinity. There is a contraction of space around the infinite light. The contraction (*tzimtzum*) is understood as addressing the paradox of the existence of the imperfect, finite world within the absolute oneness and perfection of God. In Creation, according to Kabbalah, the infinite ray of light entered space that contracted, forming ten concentric spheres. These spheres are the Sefirot. The geometrical model here is complex and mathematically meaningful. A simplified image is shown below.

Dante Alighieri (1265–1321), writing in the 1300s, used an almost identical model to describe paradise, purgatory, and hell in the *Divine Comedy*. As Dante traveled through the nine spheres of the world, he reached the nine spheres of angels. Beyond these spheres lies a point called Empyrean,

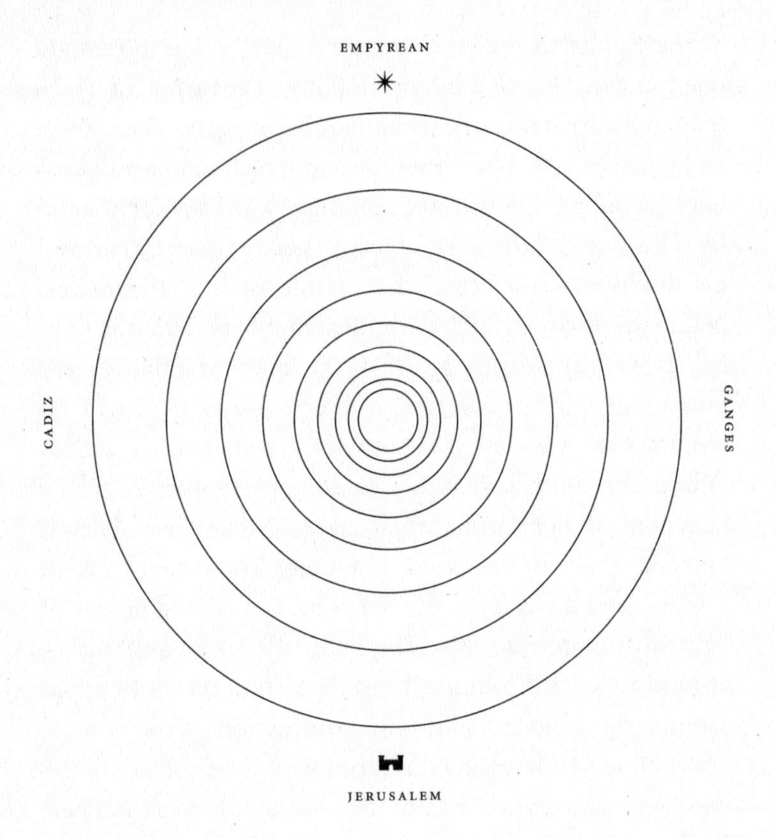

where God resides. In both the Kabbalah model and that of Dante we see that spheres lie nested with a common center, and somewhere on top of them, far away, is a *point at infinity*. In the 1800s, a great German mathematician with a keen understanding of geometry discovered a meaningful way of describing infinity using just such a sphere, now named after him: the Riemann Sphere.

Dante also developed a mystical number system. He rediscovered the Pythagoreans' tetractys, and considered ten a

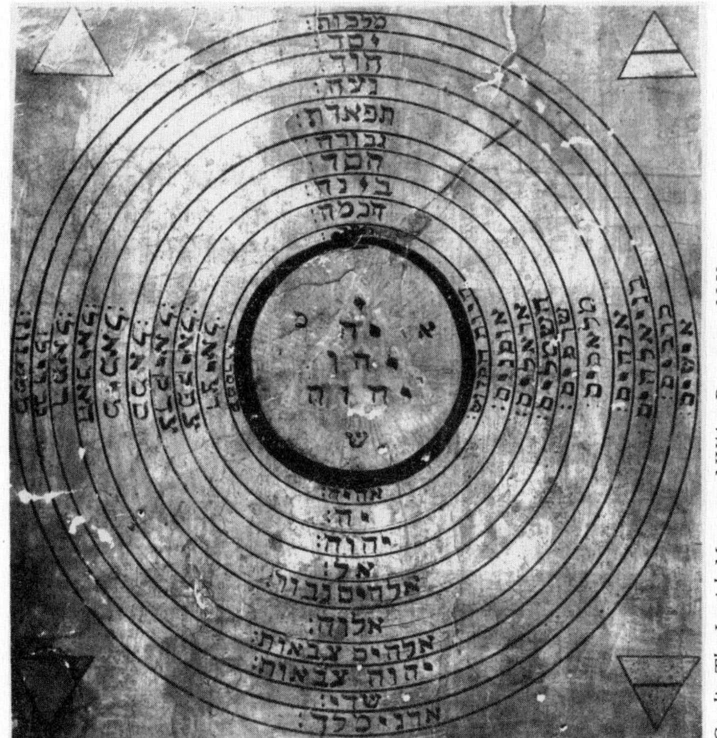

Kabbalistic plaque.

Credit: The Jewish Museum, NY/Art Resource, N.Y.

most important and sacred number. He gave Beatrice the number nine, himself a ten, and computed hidden meanings from the interactions of these numbers with letters, just as the Kabbalists did using gematria. Throughout the Middle Ages, numerology played an important role in Christian mysticism as well. The idea of infinity arrived here in discussions about angels being in two places at once, how many angels can dance on the tip of a pin, and other arguments. Kabbalistic principles were studied by Christian theologians as well. But Christian theology developed its own conceptions of infinity independent of Kabbalah.

Infinity was studied by Augustine (A.D. 354–430). In the *City of God*, Book XII, Chapter 19, Augustine wrote that "individual numbers are finite but as a class they are infinite. Does that mean that God does not know all numbers, because of their infinity? Does God's knowledge extend as far as a certain sum, and end there? No one could be insane enough to say that."

In the Middle Ages, Thomas Aquinas wrote about the concept of infinity. Aquinas was born at the Castle of Roccasecca near Naples in 1224. When he was five, his parents sent him to be educated at the Benedictine Abbey of Monte Cassino, northwest of Naples. Here at the famous abbey, the young boy was first introduced to Christian philosophy. When the King of Naples closed the abbey several years later, Aquinas returned to the city and when he came of age enrolled as a student at the University of Naples. He was later ordained into the Dominican order and in the following years traveled extensively throughout Europe from Paris to Cologne to Rome, studying and preaching.

Aquinas is famous for his attempts to prove the existence of God. Contemplating the idea that God was infinite, he arrived at a conclusion that seemed paradoxical. Aquinas asked whether the Earth was created from God's eternity. If so, there must have been an infinite number of souls in the universe, a conclusion he found problematic. Aquinas did not resolve this paradox, but developed important ideas about infinity and its nature in his attempts to do so. Before he could reach any conclusion, Aquinas died in 1274 at Fossanuova on the road from Rome to Naples.

Aquinas's concepts of infinity were later adopted by other Christian scholars. Thomas Bradwardine (c. 1290–1349), a mathematician and theologian who rose to become Archbishop of Canterbury, extended Christian studies of infinity from the discrete infinity of the number of souls or angels to the continuous infinity of geometrical figures. In his text *Tractatus de Continuo*, Bradwardine argued that continuous magnitudes are composed of an infinite number of continua of the same kind. His speculations on the nature of continuous infinity led to further work by Nicholas of Cusa (1401–1464).

Nicholas of Cusa was an ecclesiastic and mathematician who studied circles and polygons and even tried to square the circle. He became a cardinal but spent many years studying mathematical problems of antiquity. Nicholas likened the knowledge of God to a circle. He visualized human knowledge as a polygon inscribed within the circle. From these principles, Nicholas constructed a limit argument whereby as human knowledge increases, the polygon gains more and more sides, their number approaching infinity. But Nicholas concluded that no matter how much such knowledge grows, it could never reach God's knowledge in the same way that an inscribed polygon never actually becomes the circle—no matter how many sides it has.

Geometric concepts of infinity developed further during the Renaissance and were used by mathematicians, theologians, philosophers, and artists. In the 1500s, artists learned how to make use of the point at infinity in their paintings. Perspective,

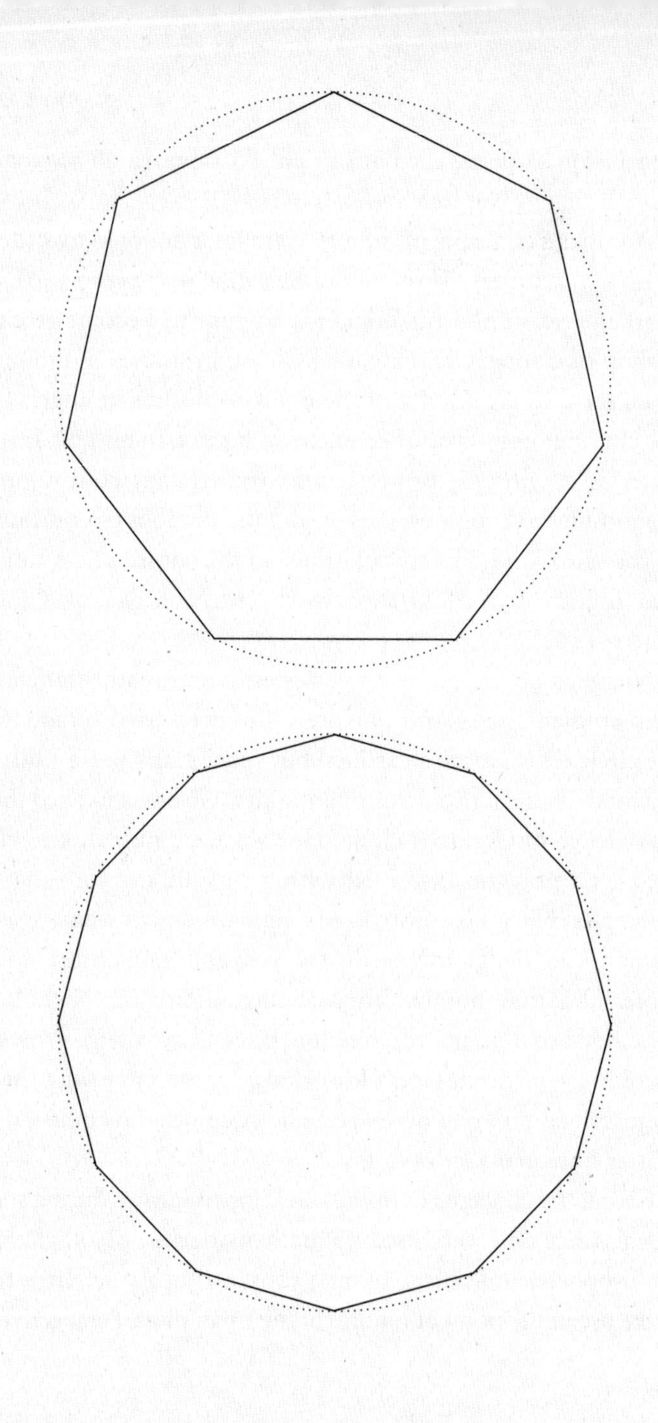

A woodcut from *La Practica della Perspecttiva* by Daniel Barbaro (Venice, 1568: C. & R. Borgominieri).

demonstrated well by paintings of the Venetian artist Giorgione and others, used a vanishing point in the center of the painting, toward which the landscape disappeared as if at great distance. The vanishing point is the point at infinity.

A point at infinity, like the vanishing point of a Renaissance painting, lies hidden behind the ten Sefirot of the Kabbalah. This point encapsulates the entire nature of the Sefirot, along with infinitely many other qualities of God hidden within the Ein Sof.

It has been said that the Kabbalah is a secret garden,

which only few can enter and survive. Even modern books about the Kabbalah warn the reader that the garden is not for everyone and must be entered and traveled through with caution. Only deep and strong personalities can benefit from the closeness of the Ein Sof. The metaphor of a garden is not accidental. The Torah, comprised of God's words, is the starting point for Jewish spiritual practice. The Kabbalist reads the Torah on four levels: Peshat (literal), Remez (homiletical), Derash (allegorical), and finally Sod (secret). The four Hebrew letters make the acronym PRDS, pronounced Pardes, which means garden. The garden of the Kabbalah thus has four levels. Through study and meditation, the Kabbalah practitioner may ascend from one level of the garden to the next, achieving increasing understanding of the Ein Sof and fulfillment at each level.

Ein Sof in Hebrew starts with aleph (א), the first letter of the alphabet. Infinity is thus denoted by aleph. The word for God in Hebrew, *Elohim*, also starts with aleph (and so does the word for "one," *Echad*). The letter aleph represents the infinite nature, and the oneness, of God.

$\aleph_3$

Galileo and Bolzano

From the early 1600s to the early 1800s, two mathematicians made deep discoveries about the nature of infinity. Their discoveries can be seen as a continuation of the keen insights of Greek mathematicians who lived two thousand years earlier. The theory of the calculus as well as other important fields of mathematics were developed and advanced during this period, and some of the greatest names in mathematics were making their impact on the field: Newton, Leibniz, Gauss, Euler, and others. None of these mathematicians, however, dared reach for infinity. The mathematicians used clever arguments in their derivations, where a quantity *approached* infinity, or where quantities approached zero. Thus, these mathematicians dealt only with *potential* infinity. None of them dared enter the secret garden.

It remained for one of the greatest scientists of all time—but not one generally associated with abstract mathematics—to discover a key property of *actual* infinity. The man was Galileo.

Galileo Galilei (1564–1642) was a unique intellect—mathematician, physicist, astronomer, humanist. Galileo's childhood was spent in Pisa, from his birth on February 15, 1564, through his family's move to Florence in 1574. He was born in Renaissance Italy—where the winds of change were blowing with new ideas and human creativity was blossoming.

His family sent the young Galileo back to Pisa to study medicine at the university. Here Galileo discovered that he was a mathematician. Without his family's knowledge, he hired a private tutor of mathematics—a former student of the legendary Italian mathematician Tartaglia. Under the tutelage of Ostilio Ricci, Galileo discovered the beautiful world of equations and geometry. Galileo found that he had a gift: the ability to view the physical world around him mathematically.

Soon, in 1583, Galileo made the first discovery of modern physics. While attending a service at the Cathedral of Pisa on a stormy day, Galileo's attention wandered away from the sermon. His eyes followed the chandelier swaying in the wind overhead. Galileo timed the rhythm of the chandelier against his pulse. He was quickly able to deduce an amazing property: longer swings and shorter ones all took the same length of time. We call Galileo's discovery the law of the isochronism of the pendulum.

Galileo continued his exciting study of mathematics. Abandoning medicine, he soon returned home to Florence without a medical doctor's degree. Jobless and facing his family's disappointment with his empty-handed return, Galileo immersed himself in mathematics and found excitement in reading ancient Greek texts about the genius Archimedes. Supposedly, he was so taken with the story of Archimedes' cry of *Eureka* upon discovering a mathematical law of nature, that he pursued the same line and made his own discoveries in hydrostatics.

By age 22, Galileo had a number of inventions and mathematical discoveries to his credit and was teaching students mathematics at several locations in Tuscany. He published his first book, *The Little Balance*, describing his discoveries extending the work of the great Archimedes. Although lacking a formal degree, Galileo's renown as a mathematician grew, and three years later, at the age of 25, he was appointed to the Chair of Mathematics at the University of Pisa.

A few years later, Galileo moved to the University of Padua, near Venice. In 1609 a Dutch delegate appeared in Padua on his way to Venice, hoping to demonstrate to the Venetians the usefulness of a new invention: the telescope. Galileo got word of the Dutch attempt and contacted a high-placed friend in Venice. He convinced the friend, Paolo Sarpi, to intervene on his behalf with the authorities, promising that he would outdo his Dutch competitor and provide the Venetians with a much better telescope.

On August 21, 1609, the entire Venetian Senate clambered up the stairs to the top of St. Mark's belltower to see how Galileo's telescope worked. Written records attest to how the Venetians were impressed with "the marvelous and effective singularity of Galileo's spy-glass." The senators were able to see clearly people walking on the island of Murano several miles away. Ships could be seen approaching the lagoon long before they were visible to the naked eye. The Venetians quickly understood the military possibilities offered by the telescope in protecting the republic from attacks by sea. The

Doge and the senators were so impressed that Venice soon bought a number of telescopes made by Galileo. The scientist was able to augment his income—and increase his prestige.

Shortly after he demonstrated the telescope to the Venetians, Galileo turned his telescope toward the skies. The physicist and mathematician now became the first modern astronomer. Galileo's amazing discoveries in the skies included the rings of Saturn, the moons of Jupiter, and the fact that the Milky Way is made of many stars. Galileo came to the conclusion that the Earth could not be the center of the universe, for with his telescope he found heavenly bodies circling another center. Galileo continued his research, and used mathematics to model the orbits of the planets. He was able to confirm to his satisfaction that the Copernican theory was correct.

Less than a year after the Venetian senate had decided to increase Galileo's pay dramatically due to his path-breaking discoveries and inventions, he suddenly chose to return to Florence. Unlike Tuscany, the Venetian Republic had asserted a great degree of independence from the Holy See in Rome. And by the time he left Padua, Galileo had published prodigiously on the Copernican theory. It would have been safer and wiser for him to remain in Padua under Venetian protection. But Galileo decided to return to Florence, where Cosimo II de Medici, the Grand Duke of Tuscany, named him Philosopher and Mathematician to the Grand Duke.

Galileo felt that his friendship with Cosimo II would protect him from any threat. But the Grand Duke's powers were not enough to save the brilliant scientist from his detractors—especially when these enemies were powerful people who could enlist to their cause the greatest force in Europe: the Inquisition.

Full of confidence in the importance of his scientific work and the truth of the philosophy that supported it, Galileo set out for Rome in 1615. In Rome, Galileo was received with great honor. He was a world-famous scientist by now, a celebrity, and he was entertained accordingly. The way he was treated by everyone from the Tuscan ambassador to officials of the Church lulled Galileo into a false sense of security. He became confident that his enemies couldn't touch him, and that the authorities would embrace his theories by the strength of his own conviction. Unbeknownst to him, his confident attitude helped set the trap that ensnared him seventeen years later.

Galileo approached the Church, hoping to win its full support for his views. He was given an audience with the influential Cardinal Bellarmine. Later, in 1616, the pope assigned Bellarmine the task of investigating whether Galileo's views as expressed in public and in his writings constituted a challenge to Church doctrine. Cardinal Bellarmine pointed to Psalm 19, which said that the Sun was in motion, not the Earth. He then interpreted Galileo's writings as a direct challenge to Scripture, and issued a mildly-worded written response. A copy of Bellarmine's letter was handed to Galileo, and the original was supposedly placed in his file in the Vatican.

Ironically, Galileo misinterpreted Bellarmine's action of 1616 and was emboldened to continue his discourse on the heliocentric theory, naively believing that the Church was about to accept the Copernican view of the world. But the copy of Bellarmine's report that was given to Galileo was not identical with the actual report placed in his file. Shortly after Galileo returned to Florence confident in the course he had chosen, Cardinal Bellarmine died. There would now be no witness to the actual interaction that took place between the two men, and no one would be able to authenticate the validity of the document in Galileo's possession against the false copy in his Inquisition file.

In 1629, Galileo put down his thoughts in a book, *Dialogue Concerning the Two Chief World Systems*, which became an overnight success. The book was a dialectical argument about the Copernican theory. Galileo sincerely believed that he was being fair to both sides of the argument. The dialogue in the book is a discussion among three individuals. Two of them are named after Galileo's friends and hold the "right" views about the Sun and the Earth. The third discussant is called Simplicius and holds the views of the Church.

The pope was initially favorably disposed toward Galileo. But as soon as Galileo published the *Dialogue*, Galileo's enemies pounced on the opportunity for which they had been waiting. Through connivance they were able to reach the pope and to convince him that Simplicius in Galileo's work was none other than the pope himself, and that the book mocked the pontiff. Their scheme worked, and in August 1632, the Vatican ordered Galileo's Florentine publisher to suspend all sales of the *Dialogue*. At the same time, the irate pontiff appointed his nephew, Cardinal Francesco Barberini, to form a committee to investigate Galileo's book.

Soon afterwards, Galileo was ordered to Rome to answer

the charges against him. The aging scientist, in poor health, asked for a postponement. His request was turned down and he was told to present himself at the Vatican within sixty days. Everyone except Galileo understood the severity of his predicament. The grand duke of Tuscany tried to intervene on behalf of his celebrated mathematician, but failed. The Venetian Republic offered to protect Galileo from the Inquisition if he would return to Venetian territory. Galileo politely turned down the offer. He was still confident he could win his case against the Church. On February 13, 1633, Galileo arrived in Rome to face trial by the Inquisition. He was made to kneel and recant his theories under threat of torture, even if he did mutter a defiant "E pur si muove," as some have suggested. In return, his death sentence was commuted to house arrest in Florence for the rest of his life.

While under house arrest, Galileo could no longer travel to conduct his famous physical experiments. He was required to stay home, his nun daughter reciting for him the daily Hail Marys the Inquisition prescribed as part of the agreement to commute his death sentence for daring to say the Earth was not the center of the universe. In 1992, at the 350th anniversary of Galileo's death, Pope John Paul II finally apologized to Galileo for his treatment by the Inquisition. For pure mathematics, however, and for our understanding of the concept of infinity, Galileo's house arrest was a blessing in disguise.

It was during this long and sad period of confinement to his home and beautiful gardens that Galileo wrote a treatise, *On Two New Sciences* (1638), in which he discussed various philosophical and mathematical ideas by way of a complicated dialogue. The intelligent voice in these dialogues is Salviati. His opponent is again named Simplicius, the simpleton. This publication was Galileo's thinly veiled revenge on the Inquisition, putting the Inquisitors' views in the mouth of Simplicius.

Salviati explains to Simplicius various aspects of infinity. He starts with the kind of infinity that was well-understood and used by the ancients as well as Renaissance and later mathematicians: the potential infinity of limits. Through Salviati, Galileo explains the division of a circle into "infinitely many" infinitesimally small triangles. He argues that by bending a line segment into the shape of a circle one has "reduced to actuality that infinite number of parts into which, while it was straight, were contained in it only potentially." Thus the circle, he continues, is a polygon with an infinite number of sides.

Galileo's argument, which harks back to Eudoxus and Archimedes and their method for deriving the areas and volumes of curved surfaces and bodies, was also used by another champion of astronomy. Johann Kepler (1571–1630) derived mathematically the laws of motion of planets around the Sun. Kepler's Laws are used today in space exploration as well as in astronomy. By using ingenious mathematical methods, Kepler was able to discover and express with equations the exact laws of planetary motion. In 1609 he announced the first two laws: Planets move about the Sun in elliptical orbits with the Sun at one focus; and the line joining a planet to the Sun sweeps equal areas in equal times. In deriving these laws, Kepler made extensive use of potential infinity. He divided the areas of the ellipses into very many "infinitesimal" triangles, then computed their areas and was able to see what the limit of the total sum of areas would be as the number of triangles increased toward infinity. In 1612, Kepler was even able to apply his method to finding volumes of wine bottles in response to the great demand for wine, 1612 having been an excellent year.

Further on in his treatise *On Two New Sciences*, Galileo goes the extra step—the big leap from *potential infinity*, used by the ancients as well as his contemporaries, to *actual infinity*, which only the Kabbalists had dared approach before him. Salviati sets up a one-to-one correspondence between all the integers and all the squares of integers and says "We must conclude that there are as many squares as there are numbers." Thus an infinite set, the set of all whole numbers, is shown to be "equal in number" to the set of all squares of whole numbers, which is a proper subset of the set of whole numbers. How can this be possible?

To understand what Galileo discovered we must first define our way of counting off things. How do we count? What is that action of "counting" things? Let's analyze it carefully.

To count the circles above, we (unconsciously) set up a one-to-one correspondence between the integers—always starting with one and moving up from there—and the items to be counted. Thus we assign the first circle the number one, the second the number two, and to the third, the last circle we have, we assign the number three. Since there are no more circles to be associated with numbers, and since three is the largest number we have associated with a circle, we know definitively that there are three circles. With only three items, this is, of course, a trivial matter; but try this with 30 or more items. Counting the items means associating with each item one and only one integer, in increasing order, until the last one has been paired off with an integer. That last number assigned is the total number of items in the group.

With any *finite* number of elements, there are no problems or apparent paradoxes. We can count (if we have enough time) as many items as there are in any set. The same counting principle works for infinite sets as well. To know "how many" elements there are in an infinite set, we still go ahead and assign an integer to each of the elements and then see how far we can go. So what Galileo did was to try to "count" all the square numbers. He assigned the number one, the first number we use in all counting, to the first square number, which is also 1. Then he assigned the next square number, 4, to the second integer, 2, as we always count. He assigned 9, the next square, to 3, the next integer, and so on, ad infinitum.

Using this normal "counting" procedure as we would with any finite set of things, Galileo found that infinite sets are very different from their finite counterparts: An infinite set can be shown to have "the same number of elements" as a proper subset of itself. Galileo noticed that associating with each number its square, "counting" the squares, there are as many squares as there are whole numbers: 1→1, 2→4, 3→9, 4→16, and so on.

But what about numbers that aren't squares, where are they in this counting assignment? This seems paradoxical. And yet it's true, every number can be put into a one-to-one relationship with its square. In a sense, there are as many numbers as there are squares. This phenomenon holds true only because both sets are infinite. When Salviati in Galileo's dialogue concluded correctly that the number of squares is not less than the number of integers, Galileo couldn't quite make him say that the two sets are of *equal* number. This was too much for him. He was shocked by the discovery that while there were (infinitely many) numbers left over—all the non-squares—each square had already been associated with an integer. Galileo had thus discovered the key property of infinite sets: An infinite set can be "equal" in number of elements to a smaller subset of itself—a set included as a smaller part of the original set. Infinity is an intimidating concept—one where our everyday intuition no longer serves to guide us. Galileo stopped there, even though he had intended to write a book about infinity. Apparently, the power of the infinite was enough to deter him from this project.

The best way to see that an infinite set can be put into a one-to-one correspondence with a proper subset of itself is to consider the *infinite hotel*. The Infinite Hotel is often called Hilbert's Hotel, in honor of the great German mathematician David Hilbert (1862–1943), who liked to tell the story. The infinite hotel has infinitely many rooms. Unfortunately, when you arrive at the hotel, the manager tells you that all the rooms are full. There is no vacancy.

"But you have *infinitely* many rooms, don't you?" you persist. "Yes, that's true," says the manager, "but all our infinitely many rooms are full. There isn't *one* of them that is vacant." You scratch your head. Infinitely many rooms; all are full. Then you have an idea: "Why don't you do this," you suggest to the manager: "Move the person in room 1 to room 2. Then move the person in room 2 to room 3, and the one in room 3 to room 4, the one in room 4 to room 5, and so on to infinity. Since you have infinitely many rooms, you can continue moving all your guests, and room 1 is now available for me."

You finally have a room in this infinitely-full hotel. In fact, you can even secure *infinitely* many empty rooms, by doing what Salviati did in Galileo's book: send the person in room 2 to room 4, the person in room 3 to room 9, the one in room 4 to room 16, and so on. You will now have infinitely many available rooms. Infinity can play weird tricks on our human imagination. This is only one of the surprises infinity has in store for us.

Galileo was the first person in history to have touched actual infinity and survived the ordeal—at least for a short time. In 1642, only a few years after his frightful encounter with the Inquisition, he died, a broken man by all accounts. Galileo understood a crucial, counter-intuitive fact about infinity: that an infinite set can be in a sense "equal" to a part of itself. Perhaps this is something the Kabbalists had in mind when they said the ten Sefirot are a part of the infinity of God. If God is infinite, certainly extracting ten elements still leaves an infinite set. The ten Sefirot are like ten rooms vacated in the infinite hotel.

At any rate, Galileo only addressed the *discrete* form of infinity, one that, while infinite, could still be *counted*. Today we call such infinite but countable sets, such as the set of all integers or all square integers, *countably infinite sets*. Some mathematicians use the term *denumerable* sets for the same thing. Going beyond the countable sets to the continuum, touched upon by the ancient Greeks in their study of geometry and the irrational numbers that so troubled the Pythagoreans, was to be the work of another mathematician.

Bernhard Bolzano (1781–1848) was a Czech priest who was shunned by his church because he held progressive views on theology. Rejected by the clergy, Bolzano did what Galileo did after his trial: he turned his attention to mathematics and the concept of infinity. In 1850, two years after his death, Bolzano's book, *Paradoxien des Unendlichen* ("Paradoxes of Infinity"), was published. Unfortunately, the book and the groundbreaking ideas it contained received little attention from mathematicians at the time.

Bernhard Bolzano was born on October 5, 1781, in the Old Town of Prague. His father was an art dealer who had migrated there from Italy. His mother, born Cecilia Maurer, exerted much influence on the intellectual development of her son Bernhard while raising eleven other children, ten of whom died young. Bernhard was a frail child who was often sick and suffered from poor vision and hearing problems which plagued him throughout his life.

Bolzano spent five years at the classical academy of the Piarists, where he did not distinguish himself in any way and found philosophy and mathematics to be especially difficult. Ironically, these were the two subjects in which he would make his name. In 1796, Bolzano enrolled at the University of Prague. Books were scarce and expensive, and so Bolzano found himself working much harder to learn. Bolzano was attracted to the works of the mathematicians of ancient Greece, especially Eudoxus of Cnidus. The work of the Greek explorer of infinity and infinitesimal quantities brought Bolzano to the study of the infinite. He also studied Euclid's geometry and works by later mathematicians, among them Euler and Lagrange. In 1817, Bolzano made an important mathematical discovery. He found a function that was continuous but not differentiable. The same discovery would be made decades later by Weierstrass who would get the credit for it while Bolzano's work remained unknown.

In 1805, Bolzano was ordained a priest and nominated to the chair of the department of the philosophy of religion at the University of Prague. Bolzano had wanted the position for several years but had been passed over for promotion by lesser-qualified but better-connected individuals. He finally achieved academic status as well as a post that would allow him to develop intellectually and to educate young minds in philosophy, religion, and mathematics. But Bolzano's tenure at the university was short-lived. A mere decade and a half after his installment as chair, Bolzano was summarily fired and stripped of his priestly rank. His story, with its ingredients of intrigue and religious intolerance, was not unlike that of Galileo.

Like all institutions of higher learning in Austro-Hungary, the University of Prague was governed from Vienna. And since the Austro-Hungarian Empire did not separate the affairs of church from state, Bolzano's appointment and the evaluation of his performance included both religious and secular elements, inexorably intertwined. Bolzano was a gifted teacher who taught both religious disciplines and mathematics, but he also was required as chairman to give sermons and other lectures on issues of social value. During one of these public lectures given by Bolzano, something happened, which, almost two centuries later, is still not clearly understood and has been the subject of controversy and debate.

Like Galileo two centuries before him, Bolzano managed to antagonize an important functionary. One B. Frint had written a textbook which he had hoped would be used by Bolzano in his courses. But Bolzano, in his new position, resisted the pressure and did not adopt the book. Frint successfully turned people against the new chair of the philosophy of religion department. The slow but systematic case against Bolzano was built in a series of state papers documenting what officials considered objectionable elements in Bolzano's sermons. The most offensive infraction was Bolzano's preaching peace to the students of his university. He said that in a few decades war would be considered an unacceptable way of solving problems between nations. War would become as unpopular as duelling had become over time in Austro-Hungary, he asserted. When the first attacks on him occurred in 1808, Bolzano had the support of the Archbishop of Prague, and this helped him evade any serious consequences.

The conflict dragged on, and ten years later, on March 31, 1818, Bolzano wrote a lengthy formal reply to all the charges against him. With respect to the Frint book, the text was too expensive for students, and it was incomplete, Bolzano stated. By May, a reply came from Vienna, concentrating on the part of the complaint addressing his sermons. It was an imperial order that Bolzano be dismissed and that he be reprimanded for his sermons concerning war. After an appeal process that ended in his favor, a final order came from the emperor in 1821 stated that Bolzano was not to be given any teaching or religious duties. Again the archbishop interceded for his friend, and three more years passed. But on the last day of 1824, a ceremonial act of dismissal took place. Bolzano had to go home. Still, he received a good pension, and at age 43 could start a new life.

Bolzano stayed for twenty summers at the estate of a wealthy dowager, until her death in 1842. The winters he spent with his sole surviving brother in Prague. It was during this peaceful period of exile from the university, alternately spent in lavish gardens in the country and in the stimulating milieu of the city of Prague, that Bolzano began to contemplate the nature of infinity.

After he had to leave the old lady's estate in 1842, Bolzano spent most of his time in Prague. He made regular visits to the village of Melnik, at the confluence of the Vltava and Labe rivers, where he discussed the paradoxes he discovered about infinity with his friend Fr. Prihonsky. After Bolzano's death on December 18, 1848, Prihonsky collected Bolzano's discoveries about infinity and edited them for publication in *Paradoxes of Infinity*. To Bolzano, in all of God's creation there was no lowest and no highest degree of being. He believed that eternity was time stretching to infinity in both directions. Assertions about infinity arising from the contemplation of God and time brought Bolzano to an understanding of mathematical infinity, and to the discovery of its paradoxical nature.

Bolzano began by addressing Galileo's paradox about countably infinite sets. He then asked whether a similar property of infinity might be exhibited by the dense numbers of the continuum. Here, he found that the same property did indeed apply. Bolzano looked at two intervals of numbers: all the numbers between zero and one, and all the numbers between zero and two. By an ingenious use of the concept of a *function*, Bolzano was able to establish the same one-to-one correspondence that Galileo had used in the discrete world of integers, here for two continua of numbers.

Here is what Bolzano did. He looked at a very simple mathematical function, $y=2x$. He let this function act on all the numbers in the domain space: all the numbers between zero and one. For each of these numbers, the function $y=2x$ assigns a unique number in the range space: the space of all numbers between zero and two. For example, the number 0.5, which lives in the domain between 0 and 1, is now assigned a value in the range (0 to 2) given by: $y=2x=2(0.5)=1$. In the same way, *every* real number (real number means a rational number or an irrational number) between 0 and 1 is assigned a unique companion between 0 and 2. Therefore, Bolzano concluded, there are *as many* numbers between 0 and 1 as there are in the interval 0 to 2, which has twice the length of the 0 to 1 interval. The function and the correspondence are shown below.

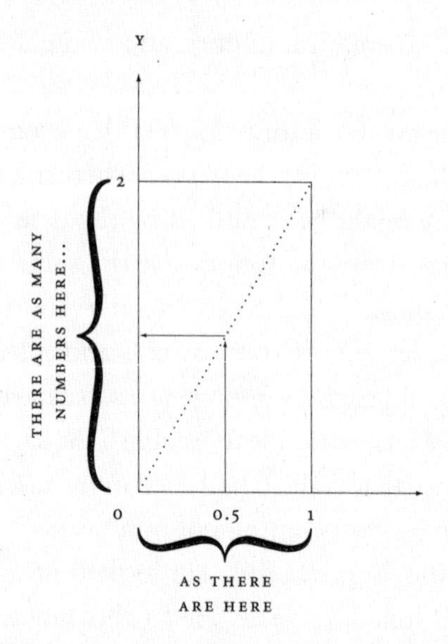

So here again, a mysterious and perplexing property of infinity bared itself: a closed interval of numbers (that is, one containing its endpoints) has *just as many numbers* as any other closed interval of numbers, no matter how large. This can be said because the function $y=2x$ was chosen arbitrarily. Had Bolzano chosen the function $y=78x$, he would have demonstrated that there are just as many numbers between 0 and 1 as there are between 0 and 78. (Both sets of numbers are infinite, but there are as many numbers in one set as there are in the other.) Other changes in the function can be used to show that the domain space need not be 0 to 1, but can be any closed interval of numbers.

Bolzano made many other contributions to mathematics. One of them is a famous result in mathematical analysis called the Bolzano-Weierstrass property. Bolzano derived and proved the property, but as with other work he had done in mathematics, he received virtually no recognition for it in his lifetime. The German mathematician Karl Weierstrass eventually rediscovered Bolzano's idea and brought it to the attention of the mathematical community. A space is said to have the Bolzano-Weierstrass property if every infinite sequence in a subset of the space has a limit point within the space. A good example of sequences and limits is given by the function $1/n$. Consider the sequence of points $1/n$ for $n=1, 2, 3, \ldots$ and so on to infinity. This infinite sequence converges to the limit point zero (because $1/n$ gets smaller and smaller as n increases: 1/2, 1/3, 1/4, 1/5, 1/6, . . . numbers getting smaller and smaller and approaching the limit zero as n goes to infinity). The Bolzano-Weierstrass Theorem says that infinite sequences in a bounded space contain limit points.

ℵ4

Berlin

By the late 1800s, the facts about infinity described earlier were known, but few mathematicians paid them any attention. At that time, there were three great centers of mathematics in Europe. These were the departments of mathematics at the universities of Paris, Milan, and Berlin.

Berlin was the center of the world for all German-speaking mathematicians. The department of mathematics at Berlin was full of world-renowned stars in the field. In fact, during the period from 1860 to the start of World War I, Berlin was the undisputed leader in world mathematics.

German mathematics began its climb to world fame at the turn of the nineteenth century with the work of the great Gauss. Carl Friedrich Gauss was a child prodigy who at a very young age had already derived many important results in mathematics decades before they were even considered by other mathematicians. Gauss taught at the University of Göttingen, but his disciples helped found the school of mathematics at the University of Berlin. Among them was

Peter G. L. Dirichlet (1805–1859), Gauss's most devoted student. Dirichlet was known always to carry with him his master's book, *Disquisitiones*, containing Gauss's great mathematical ideas. Dirichlet thus brought Gauss's groundbreaking mathematical discoveries with him to Berlin, where, thanks to Dirichlet, modern mathematical analysis was born.

Gifted mathematicians there included Bernhard Riemann (1826–1866), who, besides doing innovative work in geometry, also made the idea of the integral more rigorous. His work in geometry led Riemann to consider the problem of infinity. The infinitude of straight lines is implied in Euclid's second postulate. Riemann argued that Euclid's lines could also be interpreted as unbounded and yet not infinite. A great circle on a sphere can be interpreted as a line that is unbounded but finite. Riemann's mathematical foresight was so keen that the British astronomer Sir Arthur Eddington would later say: "A geometer like Riemann might almost have foreseen the more important features of the actual world." Riemann began to show signs of mathematical genius at the age of six, when he was not only able to solve any arithmetical problem presented to him but to propose new problems to his baffled teachers. When he was ten, Riemann was given lessons in mathematics by a professional teacher who found that Riemann's solutions to problems were better than his own. At the age of fourteen, Riemann invented a perpetual calendar, which he gave as a present to his parents.

Riemann was a very shy boy, and he tried to overcome this shyness by compulsive preparation for every public speaking occasion. As an adolescent, he became a perfectionist who would not let any piece of work be seen by others until it was best. This propensity to avoid surprises played an important role in his academic life.

In 1846, the nineteen-year-old Riemann enrolled at the famous University of Göttingen to study theology. His decision was motivated by a desire to please his father, who wanted his son to follow in his footsteps to the clergy. But soon the young Riemann became attracted to the mathematical teachings of Gauss. With his father's grudging permission, Riemann changed course to mathematics. After a year at Göttingen, Riemann transferred to the University of Berlin, where he received an excellent mathematical education, his mind molded further by the renowned mathematicians Jacobi, Steiner, Dirichlet, Eisenstein, and others. In 1849 Riemann returned to the University of Göttingen to work on his doctorate under Gauss. Riemann made important contributions to geometry and went on to do work in number theory. In 1850, after considering problems in many areas of mathematics as well as physics, Riemann came to a deep philosophical conviction that a complete mathematical theory must be established, which would take the elementary laws governing points and transform them to the great generality of the *plenum* (by which he meant continuously-filled space).

Riemann also understood that the property of a surface

that he needed to understand and "capture" was the notion of a *distance* (also called a *metric*, a word that comes from the same root as the unit of distance, the *meter*). In the "flat" Euclidean space, the shortest distance between two points is the hypotenuse AC of the right triangle ABC if the distance along the x direction is BC and the distance along the y direction is AB. This goes back to the theorem of Pythagoras, discovered in the sixth century B.C., which led to the realization that some numbers, such as the square root of two, are irrational.

Riemann generalized the metric of Pythagoras to more complex spaces. His contributions to our story are numerous. First, the Riemann Integral of calculus is defined as an infinite sum of integrals of step functions. Such infinite sums became the starting points for the study of infinity by Georg Cantor. Second, Riemann's metric is a generalization of the Pythagorean formula, which 2,500 years earlier led the Pythagoreans to discover the irrational numbers. Finally, Riemann's work on geometry touched directly on the concept of infinity in its treatment of Euclid's theory of space.

Riemann extended Bolzano's principle showing that the number of the infinitely many points on the interval between zero and one is the same as the number of infinitely many points between zero and two. Riemann discovered what we now call the Riemann Sphere. This sphere shows how the infinitely many points on a plane can be made *compact* by adding a "point at infinity" to the sphere. This is demonstrated below.

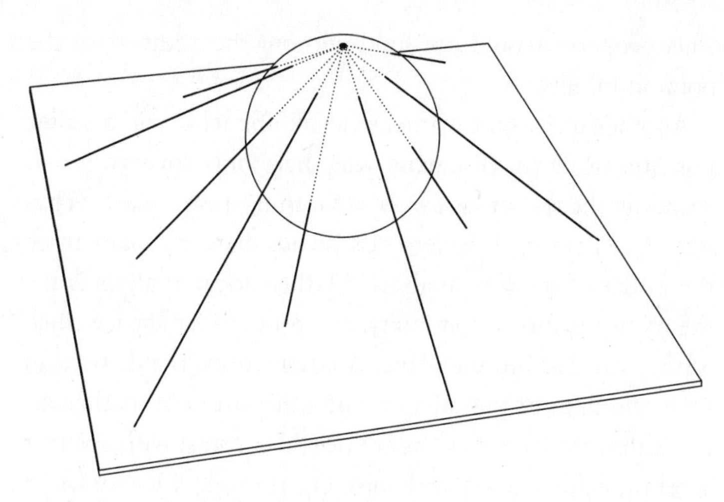

The concentric spheres of the Kabbalah, as well as Dante's spheres of the *Divine Comedy*, both leading to an infinite point representing God, are equivalents of the Riemann Sphere. As in Bolzano's transformation of one interval into another, the Riemann Sphere represents the two-dimensional plane. But here we have an added advantage: the north pole acts as a point at infinity, a point to which all the lines and points on the plane tend as we go infinitely far in *any* direction in the plane. Thus instead of the two concepts, positive infinity and negative infinity, at the two "ends" of the real one-dimensional line, here we have a plane that—using the sphere as a model—curves upon itself, tending to *one* point at infinity from any direction as one goes far enough (all directions eventually lead "north"). The plane thus becomes *compact.* (The plane includes its limit point; it is closed and bounded. All sequences of points converge within the space.)

This property would not hold without the addition of the point at infinity.

Another important mathematician, one who held a senior position in Berlin (Riemann was there only for two years, spending the rest of his career at Göttingen) was Karl Weierstrass (1815–1897). Weierstrass is considered by many to be the father of modern analysis. Mathematical analysis is the theory of functions, continuity, and properties of spaces such as the real line and the plane. Analysis forms the theoretical underpinnings of the calculus and other areas of mathematics dealing with continuous entities (in contrast with abstract algebra, which deals with discrete things). Weierstrass, a decade older than Riemann, embarked in the 1850s on a thorough study of irrational numbers and continuity. This study started where it was left 2,300 years earlier by Eudoxus, building mathematical ideas on the foundation laid by Zeno.

Karl Wilhelm Theodor Weierstrass was born in 1815, the year of Waterloo, in Ostenfelde, in the district of Münster, Germany. He was the oldest of four children of Wilhelm and Theodora Weierstrass. The father was a customs official paid by the French. The French were still dominant in Europe, but beginning their decline as German nationalism was rising. In mathematics, too, the Germans were becoming prominent, although many French mathematicians were still making great advances in the field. When Karl was 11, his mother died and his father remarried. Apparently the stepmother did not have much interest in the intellectual development of Karl and his brother and sisters. The autocratic father was in a hurry to get his firstborn son a practical profession. At fourteen, Karl entered the Catholic Gymnasium at Paderborn, after his family had moved to Westphalia. He graduated from the gymnasium at nineteen, having excelled in his studies and received many prizes: in German, Latin, Greek, and mathematics. Unlike many mathematicians, Weierstrass was not gifted in music and never showed interest in listening to music. As an adult he would fall asleep at concerts or the opera whenever his family dragged him to a performance.

At fifteen, Karl Weierstrass worked part-time as a bookkeeper for a family acquaintance. He amazed the adults with the unusual skill he had with numbers. This led his father to decide that his son should train for a government accounting career, and he sent him at nineteen to study bookkeeping and law at the University of Bonn. But Karl, whose real talents lay in abstract mathematical thinking, was bored with the practical subjects he was required to study at the University. He spent all his time fencing, socializing, and drinking beer. A large man, he was also quick and agile, which made him an excellent fencer. It was said that he never lost a match. The young Weierstrass returned home four years later, without a degree.

What his disappointed father and siblings did not understand was that Karl had returned home with more than the unobtained degree. He had acquired the compassion, patience, and social skills that would in time help make him the best mathematics teacher of his generation. The family was devastated nonetheless: they were so deeply disappointed in Karl that they spoke of him as if he were dead. The brightest child, the one who had held the most promise of success, had lost four years of his life and wasted some of the family's limited resources.

After several weeks, a family friend proposed a solution:

Karl should study to be a teacher. He could prepare himself for a state teachers' examination at the neighboring Academy of Münster. The young man begged for a second chance, and the father agreed to give it to him. Karl enrolled at the academy. In 1839, when he was 24, Karl Weierstrass matriculated from the teaching academy and began to study for the state examinations, which would allow him to become a high school teacher.

Weierstrass did not major in pure mathematics at Münster, but he took courses offered by an inspiring mathematics professor, Christof Gudermann (1798–1852). In these courses, Weierstrass absorbed deep mathematical ideas that made him think in new ways. Eventually, Gudermann's ideas led Weierstrass to the novel mathematical theories on which he would make his name.

Gudermann studied elliptic functions, special functions that had been studied by many other mathematicians, from a new direction. Gudermann based his entire approach to function theory on power series expansions. This idea entailed approximating a given function by a *theoretically-infinite* sum of special functions. Such power series expansions had been used by Brook Taylor (1685–1731), Colin Maclaurin (1698–1746), James Stirling (1692–1770), and others. But Gudermann was using the potential-infinity idea of the ancient Greeks in assuming that a function behaves *in the limit* as would an "infinite" sum of power terms. Gudermann did not achieve much on his own by using the new approach. However, his brilliant idea of using power series to study complicated functions eventually became the key element in Weierstrass's life work—the development of mathematical analysis.

Gudermann started his course at the academy with thirteen students, but since teachers-in-training had little interest in pure mathematics, within a week the class size shrank to one student: Karl Weierstrass. The student learned so much, in fact, that after he became one of the greatest mathematicians in the world, he never lost an opportunity to acknowledge his debt to his professor at the teachers' academy.

In 1841, when he was 26 years old, Weierstrass took his exams for a teacher's certificate. The candidates taking the exam had six months in which to answer three comprehensive questions. At Weierstrass's request, one of the questions was submitted by Professor Gudermann. This problem was not a typical examination question, to which the professor knows the answer and simply checks the student's work. Gudermann's problem was a difficult mathematical conundrum, going far beyond the realm of questions asked of future teachers. Gudermann asked Weierstrass to derive the power-series expansion of elliptic functions—a hitherto unsolved problem in mathematics.

In his final report, Gudermann wrote about Weierstrass's performance on the test: "This problem, which in general would be far too difficult for a young analyst, was set at the candidate's request with the consent of the commission." Gudermann then arranged for his star student to get a special certificate indicating that he had solved a previously-unsolved problem in mathematics and in doing so had made an original contribution to the field. He went on to say that the advance achieved by Weierstrass was so significant that for the sake of science he should *not* become a schoolteacher but instead should be allowed to join the faculty of an academic institution. Gudermann's remarks were struck out of the official certificate. Weierstrass became a teacher and remained a schoolteacher, teaching German, geography, calligraphy, and other basic skills to young boys, until he was forty years old.

For fifteen years, Weierstrass taught school at small German villages, where good books were not available and where intellectual pursuits and stimulating conversations were hard to find. Throughout this period, he worked alone at night developing the modern theory of mathematical analysis as we know it today. His compensation was so low that he could not afford to pay for stamps to mail his research papers to academic journals, and so he did not publish. He worked in virtual isolation from the world's mathematical community.

Weierstrass's first publication was a paper on mathematics he wrote for a school's bulletin that from time to time included works by the teachers. But in 1854, *Crelle's Journal* (the German *Journal for Pure and Applied Mathematics*), one of the leading mathematical journals in the world, published a paper by Weierstrass, who had finally sent his important work out for publication. Overnight, the obscure schoolteacher became a mathematical celebrity. What struck the mathematicians in Berlin was not only the monumental nature of the mathematical developments achieved by an unknown teacher from a remote village, but the fact that there were no preliminary results to herald the discoveries. Weierstrass had worked patiently and not published earlier results leading to his masterpiece, as others might have done. He waited until his work was completed, and then published it in full. The result was swift—Weierstrass was offered a professorship at the University of Berlin. To support the appointment, an honorary doctor's degree in mathematics by another institution was conferred upon the bewildered schoolteacher.

In Berlin, Weierstrass continued his study of functions. His lectures on the theory of functions at the University of Berlin were so popular that his lecture hall became the place to be for mathematicians who wanted to learn the subject. Weierstrass developed Gudermann's notion of a power series expansion of a function. A power series is an infinite sum of functions. We can't add infinitely many terms, but as we add more and more terms, the finite sums get closer and closer to the function of interest. That is, the series of functions converges to the required function. Here, the idea of infinity is crucial, since the sum of functions "becomes" the desired function only when one "reaches infinity." This is the backbone of the modern approach to the theory of functions. Weierstrass also developed the notion of closeness used in studies of the continuity of functions, thus making use of infinity in the tradition of Zeno and Eudoxus. Additionally, the argument of the convergence of functions led to a rigorous definition of irrational numbers as limits of sequences of rational numbers. Thus, for example, the sequence of rational numbers: 1, 14/10, 141/100, 1414/1000, . . . converges to the *irrational* number the square root of two.

Weierstrass's antagonist at the University of Berlin was Leopold Kronecker (1823–1891). Kronecker came from a wealthy family of business people and didn't need to work as a mathematician to make a living. As a young man he showed such aptitude for mathematics that he was offered a professorship and gave up his brilliant business career to pursue mathematics. He considered music the greatest art—except for mathematics, which he likened to poetry. Kronecker's delta function, which indicates a property by being equal to one when the quality holds and zero otherwise, is his most famous invention. Kronecker's interests were within the theory of numbers, and he wrote a doctoral dissertation on that topic at the University of Berlin in 1845. His work was inspired by another well-known mathematician at the university, Ernst Eduard Kummer (1810–1893).

Kummer did important work in number theory and made advances in the understanding of Fermat's Last Theorem. When Kronecker returned to Berlin as a professor, he and Kummer conducted joint research. Kronecker's dissertation dealt with algebraic number fields that arise from a problem posed by Gauss. Gauss wanted to know how to divide the circumference of a circle into *n* equal arcs. This problem led to certain equations studied by Kronecker. They require an understanding of algebraic theories. The areas of algebra, where equations and their solutions are studied, is opposite, in some sense, to the areas of mathematical analysis. Algebra deals with *discrete* entities: integers, rational numbers (which are ratios of integers) and other elements that can be counted or arranged on a list. Mathematical analysis, on the other hand, deals with *continuous* entities: functions, intervals of numbers, and, therefore, also irrational numbers (which have unending and non-repeating decimal representations). Because the two fields are so different (and have remained so to this day) mathematicians who work in the two fields tend to think differently. Algebraists think in discrete terms, while analysts tend to visualize numbers and other mathematical entities on a continuum.

Given the difference between the two fields, it is not hard to imagine that Weierstrass, who was the father of modern analysis, and Kronecker, who made important contributions to algebra, would not get along. The two were different in other ways as well: Weierstrass was a large, overbearing man, while Kronecker was diminutive. People who watched the two mathematicians fight over mathematical theory and the essence of truth were always struck by the comic nature of these clashes, with the small man constantly attacking the large one like a small dog going after a St. Bernard.

Kronecker believed that "God made the integers, and all the rest is the work of man." He pursued mathematics believing it should deal only with the discrete elements of algebra, ignoring the important advances Weierstrass and his followers were making in analysis and allied fields such as geometry and topology.

Unwittingly, a brilliant young student of mathematics at Berlin would soon land in the middle of this bitter war between algebra and analysis. Within a few years, Kronecker would turn all his venom on the newcomer, making him his main victim. Kronecker would single-handedly prevent the student, by then a graduate with a Ph.D., from ever achieving a position at the University of Berlin, even though there would be no one more deserving.

Much of modern mathematical analysis is concerned with the behavior of functions. An important component of mathematical analysis is the study of *continuous* functions. A continuous function is one that moves continuously through points on the line and is not torn anywhere. A tear could occur if a function is defined to equal 0 for all values of x less than or equal to 5, and then, *suddenly* to equal 1 for all points higher than $x=5$. A function such as $y=2x$, for example, has no such *discontinuities*: tears or unexpected jumps, and is therefore a continuous function. Continuous functions have nice properties. One example is a fixed-point theorem demonstrated in the following story.

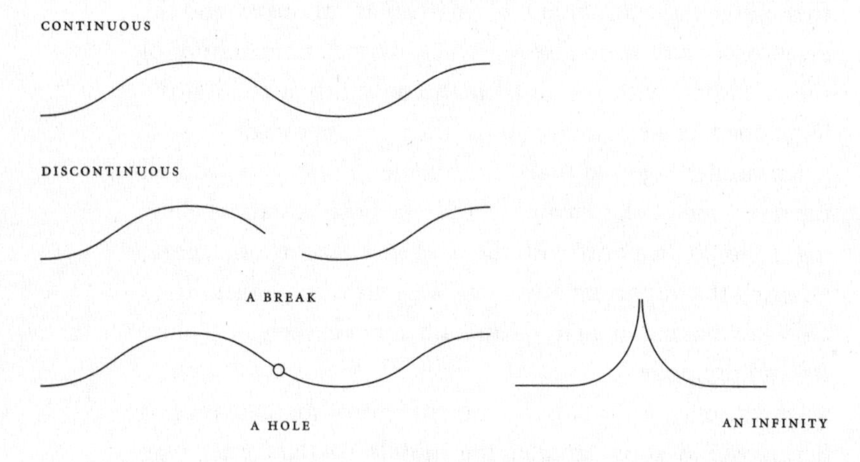

A hiker decides to climb a high mountain. He packs a bag with food and other necessities and at dawn (let's say 6 A.M.) starts up the single path toward the summit. He takes his time, stopping now and then to rest or view the scenery, even retracing his way back down a few steps to smell a flower or look at a bird on a bush. At sunset (let's say 6 P.M.), he arrives at the top of the mountain. He unpacks his belongings, pitches a tent and spends the night at the summit. Again at dawn (6 A.M.), he starts on his way down the mountain. This time, too, he takes his time, stopping wherever he wants, eating his lunch at a different place from the one he used on his ascent, and at times even going back up to look at a cave or an interesting rock formation. Again at sunset (6 P.M.), the hiker arrives at the bottom of the path from the mountain. The question is this: Is there, necessarily, a point on the path at which the hiker was at exactly the same time of day both on his ascent and on his descent from the mountain? [Answer next page.]

In the late 1800s, mathematicians started to apply their theories not only to continuous functions, but also to the harder-to-understand functions with discontinuities: functions that jump at times from one point to another rather than move smoothly from point to point. These "abnormal" functions were shown to be extremely important in such areas as the theory of integration—the field within mathematical analysis in which areas, volumes, and averages are studied. Here, discontinuous functions were used as the building blocks leading to the definition of the crucial concept of the integral. To evaluate an integral, mathematicians had to resort in some cases to the concept of *convergence*. They had to learn about the convergence of series of discontinuous functions to a continuous function for which they needed to evaluate the integral. Step functions are the basic elements that converge to a smooth function.

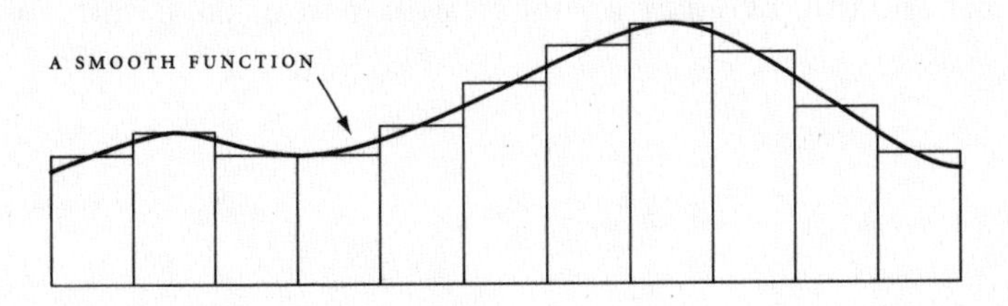

APPROXIMATING STEP-FUNCTIONS

The approximation of a continuous curve by discontinuous step functions only becomes perfect once the step functions converge to the smooth curve. This can happen when the number of step functions approaches infinity. Gauss himself believed only in "potential" infinity—something that one doesn't quite reach—an ideal, a very faraway place or number that doesn't really materialize. When we have many steps, their total area is close to that of the smooth curve and there is no need to "reach" infinity to compute the limit of the step functions. The approximation can be done to good accuracy at any finite level. This was enough for Gauss and his contemporaries. Newton and Leibniz, who invented the field called calculus two centuries earlier, were content with the idea of a potential, unreachable infinity. None had gone further.

Answer to the problem of the hiker climbing the mountain:

The graph below shows the hiker's position on the path up or down the mountain as the place on the *Y*-axis (the vertical axis), while time, from 6 A.M. to 6 P.M., is shown on the *X*-axis (the horizontal axis) for both the ascent and the descent from the mountain. Note that the only requirement for solving this problem is *continuity* of the walking function (thus the hiker can't jump from one place on the path, say a high rock, onto a lower point on the path, cutting a corner and thus avoiding having to walk part of the path).

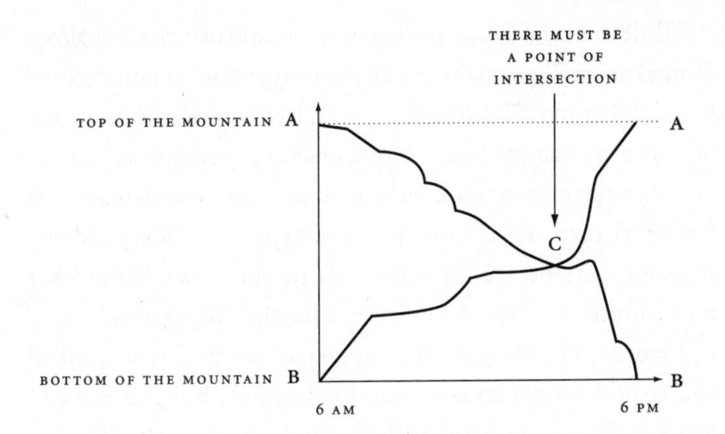

As the graph of the ascent and descent shows, no matter what the two functions look like (no matter how fast the hiker walks at any point, no matter whether or not he stops along the way or even retraces his steps), there must be a point on the path that the hiker reaches at exactly the same time both on his ascent and on his descent.

As late as the 1870s, women were not allowed to take graduate courses in mathematics at the University of Berlin. One gifted student of mathematics was not deterred. Sonja (Sophie) Kowalewski (1850-1891) was born in Moscow and by the age of fifteen began to study mathematics. She had heard of the great mathematician Weierstrass at Berlin and how he was changing the face of mathematical analysis, and decided to pursue her education under his guidance. At eighteen she married in Moscow and within a year, leaving her husband at home, traveled to Berlin and tried to meet Weierstrass. The old master sympathized with the young woman and her ambitions, remembering well the lucky breaks that brought him from a school in a small village to the center of world mathematics.

While he could not prevail on the authorities to allow Sonja to enroll as a student at the university, he volunteered to teach her himself, during his own free time. For four years, Weierstrass taught Sonja Kowalewski every Sunday at his home, and once a week they met to study mathematics at her apartment. Then Kowalewski disappeared. She suddenly decided to return to Moscow, left mathematics behind her and adopted the life of a married socialite. Weierstrass wrote her letters, inquiring about her departure and when and if she would return to continue her studies, but she did not answer. Then, as suddenly as she had disappeared, she returned to Berlin. She apologized to Weierstrass for not answering his letters and they resumed their study. Within a few years, Sonja Kowalewski became a prominent mathematician. With Weierstrass's help, Kowalewski was offered a professorship for life at the University of Stockholm, through the help of Gösta Mittag-Leffler. She received the Bordin Prize in mathematics from the French Academy of Sciences for her work on mathematical physics. Within two years, however, when she was forty-one, she died of influenza. Kowalewski did important work in mathematical analysis and, along with Weierstrass, Riemann, and others she is credited with developing the discipline and bringing it to the position it now holds within modern mathematics.

Much of what we know about the life of Georg Cantor comes from the correspondence between him and Sonja Kowalewski, as well as letters exchanged between Kowalewski and the Swedish mathematician Gösta Mittag-Leffler.

ℵ5

Squaring the Circle

Much of the work in mathematical analysis done by Weierstrass, Riemann, Kowalewski and other analysts of the time revolved around the central idea of irrational numbers. What are irrational numbers, and why are they so important?

The irrational numbers appear as though by magic as soon as we try to reconcile the line of geometry with the numbers of arithmetic so that a point on a line is viewed as a unique real number. We know that we should be able to put numbers on a straight line, to give meaning to the idea of a distance between two numbers, and to the concept of precedence of points—meaning which of two points comes first and which comes second. We should also be able to go back and forth between numbers and their representation as points on a line. If six is greater than four, then representing both numbers as points on a line segment should prove useful—we should be able to see that four is to the left of six, and to visualize the distance between the two numbers.

THE NUMBER LINE

0 2 4 6 8

On the line, we can also associate points with fractions. Between 0 and 1 are numbers such as 1/2, 1/4, 1/5, etc. Between 1 and 2 are 1 1/2, 1 1/4, and so on. Other numbers, such as 358/719 and the like—all fractions—are found easily on the number line. But the real *length*, the "meat" of the real line does not come from squeezing numbers onto it. Even with infinite condensation of *all* the fractions and integers—all the *rational* numbers—onto a segment of the number line, all we would have is a sieve riddled with infinitely many holes, but not a solid line. The real fabric of the line requires the *irrational numbers*. Without the irrational numbers, we would have an infinite collection of dots, very dense, but not solid—not a line.

Removing all the rational numbers from the number line still leaves us a line of *full-length*. This line, however, will have infinitely many holes in it. The structure of the real line is a mystery: the line is infinitely dense, infinitely condensed, with an infinitely-intricate structure. Bolzano thought that what made the continuum hold together was its property of *connectedness*, that is, the fact that any portion of the real line—any interval, as small as it may be—could not be written as the union of two disconnected open sets of numbers. (An interval of numbers that does not include its endpoints is an example of an open set.) As Cantor would teach us, the numbers on the real number line have a much more complex structure, connectedness being only one of its properties.

While the irrational numbers give the number line its fabric, the rational numbers are dense within the set of irrational numbers: as close as you wish to any irrational number are infinitely many rational ones. And vice versa: in every tiny neighborhood of a given rational number there are infinitely many irrational ones. The structure of the real number line is difficult to imagine.

The numbers are ordered: given *any* two distinct numbers, a and b, we have that either $a>b$ or $b>a$. But here is a baffling property. Given any number, there is *no next number*. If b is greater than a, then there is some distance between them. Divide that distance by 2 and add it to a and you have a new number between a and b. For example, 5.01 is greater than 5. The number 5.005 lies halfway between 5.01 and 5. I can now find a new number, one lying between 5 and 5.005, and so on. Clearly, there is no "next" number to 5. The numbers are infinitely dense, one is always greater than another, and yet there is no next number as you go from smaller to larger.

To prove that the fabric of the number line is supplied by the irrational and not the rational numbers, we use an argument similar to Zeno's paradox of never being able to leave a room. Remember that in analyzing that paradox, a property was used that the infinite series produced by always taking half the remaining distance to the door converges: $1+1/2+1/4+1/8+1/16+1/32+1/64. . . .=2$. This is an important mathematical property of the sum of a geometric series.

The rational numbers can be *enumerated,* or counted off, even though they are infinite (a property that Georg Cantor proved). The irrational numbers are so infinite that they cannot be enumerated (another property Cantor proved). Now look at all the numbers between 0 and 1. The length of the interval, when intact, is 1-0=1 unit. Now let's remove all the rational numbers. Doing so we enclose each rational number within a tiny sub-interval, like putting a tiny umbrella on top of each number. The size of the umbrella we place on top of each number decreases by half for each successive rational number. Starting with an umbrella of size ϵ (an arbitrarily tiny number, such as 0.00000001), the sum of the lengths of all these infinitely many tiny umbrellas is $\epsilon(1+1/2+1/4+1/8\ldots)=2\epsilon$. Since ϵ was arbitrarily small, the original interval between 0 and 1 has thus lost an insignificant total length, leaving it essentially at length 1 as it was when all the rational numbers were still included. We say that the rational numbers have *measure zero* within the number line. The argument above is an example of a mathematical proof.

A number is either rational or irrational and the two groups are infinitely intermingled on the line. Yet when all the rational numbers are removed, the total length of the line remains the same—there are infinitely many more irrational numbers than rational ones.

The irrational numbers themselves fall into several groups. Numbers such as the square root of two, while irrational and written with unending, nonrepeating decimal parts, can still be handled, in a sense. Such numbers are called *algebraic* numbers, because they are roots of polynomial equations with rational coefficients. The square root of 2, for example, is a root of the equation $x^2 - 2 = 0$. Since this is a polynomial with coefficient 1, the root is algebraic. Cantor proved that the set of algebraic numbers has the same size as the set of rational numbers. Non-algebraic irrational numbers are called *transcendental* numbers. All the famous irrational numbers, such as π and e, are transcendental. These are the "truly irrational" numbers—they do not take to being counted, as roots of polynomial equations (or in any other way).

The most celebrated problem in mathematics was the ancient problem of squaring the circle. This problem is essentially a statement about one transcendental number: π.

In the fifth century B.C. there lived in Athens a mathematician and philosopher named Anaxagoras (d. 428 B.C.). This was the age of Pericles, the heyday of Greek culture and achievements, and during this period many intellectuals came from all over the Greek world to live in Athens. Anaxagoras became Pericles's philosophical mentor. Anaxagoras asserted that the Sun was not a deity, but rather a huge red-hot stone in the sky, and was bigger than the entire Peloponnessus. For this heresy, Anaxagoras was arrested and sentenced to a prison term. Pericles intervened on behalf of his mentor, and eventually secured the mathematician's release.

During his period of imprisonment, however, Anaxagoras kept himself busy by trying to prove a mathematical statement. The Roman historian Plutarch described the problem Anaxagoras tried to solve while in jail, and his writings give us the first historical description of the problem of squaring the circle. Using a straightedge and compass, Anaxagoras tried to construct a square whose area is exactly that of a given circle. Thus was born the greatest mathematical problem in history, one that would capture the imaginations of mathematicians for almost two-and-a-half millennia.

This problem, together with the problem of trisecting an angle, and that of duplicating a cube, which were proposed at about the same time, set apart the mathematics of Greece from that of Babylonia and Egypt. The pursuit of solutions to these abstract problems exemplifies endeavors with no practical applications or use in technology or engineering or any other field. Anaxagoras's problem of squaring the circle was a purely intellectual pursuit.

Throughout antiquity, gifted mathematicians tried to solve these three problems. Much of Archimedes's famous study of the spiral was part of Greek efforts to find solutions. Pappus of Alexandria (c. A.D. 320) and later mathematicians also attempted to find solutions to these problems. Gregory of St. Vincent (1584–1667) wrote a book about squaring the circle and the conic sections. He used a faulty application of indivisibles, leading him to think he had achieved a solution to the age-old problem of squaring of the circle.

In 1761, a Swiss-German mathematician, Johann H. Lambert (1728–1777) presented a proof to the Berlin Academy that the number π was irrational. Lambert actually proved something more general. He showed that if x is any nonzero rational number, then $\tan x$ (the value of the tangent function of x) cannot be rational.[7] Since $\tan(\pi/4) = 1$, which is a rational number, it followed immediately that the number $\pi/4$ could not possibly be rational. Therefore, π itself could not be rational either. This proof that π was irrational did not solve the problem of squaring the circle. It could be shown that quadratic irrationalities could still be constructible by the methods of the ancient Greeks, using straightedge and compass, so—at least in principle—the problem of squaring the circle might still have held a solution.

By that time, circle-squarers had multiplied to such an extent that the Paris Academy had passed a law stating that purported solutions to the ancient problem would no longer be read by members of the academy.

Finally, in 1882, the German mathematician C. L. F. Lindemann (1852-1939) published a paper titled "On the number π." Lindemann showed that the number π could not be algebraic—it could not possibly be the solution of any polynomial equation with rational coefficients. Lindemann achieved the proof by showing that the famous equation $e^{ix} + 1 = 0$, proposed by the prolific Swiss mathematician Leonhard Euler (1707–1783), could not be satisfied by any number x if x is algebraic. Since the number π does satisfy the equation, it could not be algebraic.

In order to square the circle, one needed to be able to write π using a finite number of integers and a finite number of operations. Since, as Lindemann had showed, π could not be the root of any polynomial equation with rational coefficients, it is impossible to square the circle. Lindemann was so elated by his result, which lay to rest such an ancient problem, that he went on to propose a proof of Fermat's Last Theorem. His proof failed.

Irrational numbers that are not algebraic are transcendental. Most numbers on the number line are transcendental.

While algebraic numbers and rational numbers are infinite, the transcendental numbers are of a higher order of infinity. If you could randomly "choose" a number on the real line, the number will be transcendental with probability one. Choosing a rational number, or an algebraic one—even though there are infinitely many of them—is just too unlikely because of the preponderance of the transcendental numbers. Thus finding a rational or algebraic number when choosing a number at random from the real line has zero probability. Whether one could actually choose a number from an infinite collection of numbers is an important question which we will revisit later. The other two problems of antiquity are impossible as well.

There is an interesting way of viewing rational and irrational numbers. The array below shows the integers, 0, 1, 2, 3, 4, and so on, in two dimensions. We now draw rays emanating from the origin (0,0). The slope of a ray is a rational number if the ray—as it extends from the origin toward infinity—hits one of the points in the array. Otherwise, the slope of the ray is irrational.

How do we know that there exist lines that never hit any of the dots in the array above? (That is, how do we know that lines with irrational slopes really exist?)

Take a circle with diameter equal to one. Its circumference is then equal to π. Unfurl the circle into a line and stretch it upward from the point 1 on the horizontal axis of the array above. Connecting the point at the end of the circumference line with the origin will give you a line with slope equal to π. This ray will never intercept a point on the array.

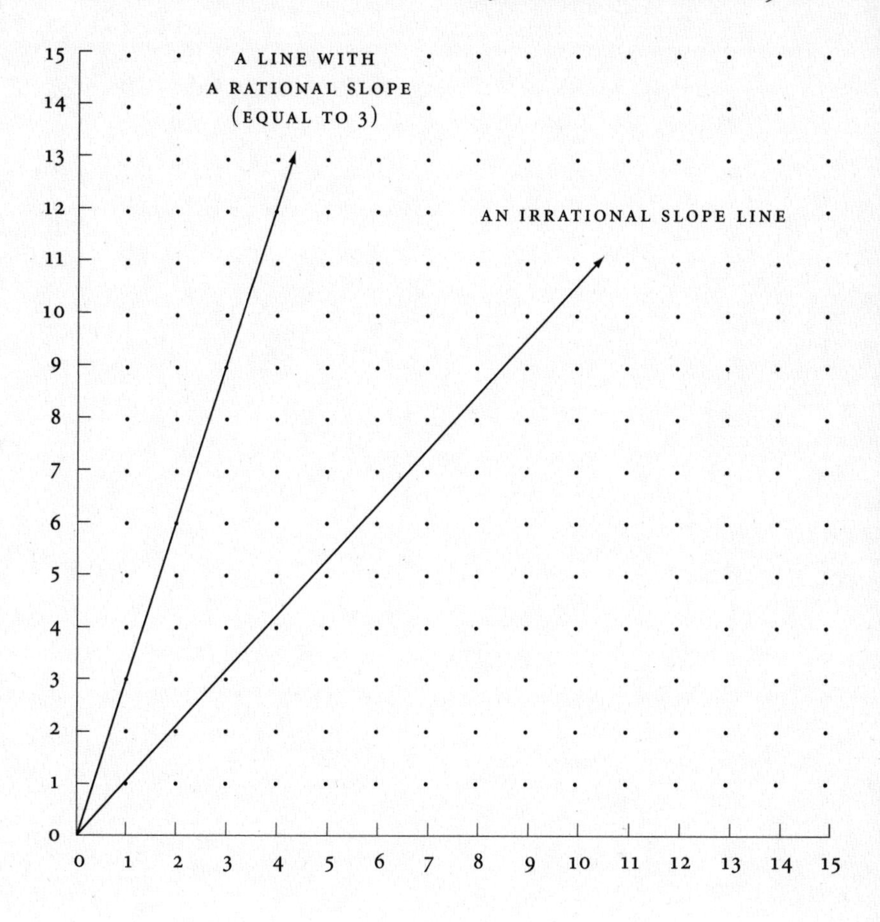
A LINE WITH
A RATIONAL SLOPE
(EQUAL TO 3)
AN IRRATIONAL SLOPE LINE
0 1 2 3 4 5 6 7 8 9 10 11 12 13 14 15
0 1 2 3 4 5 6 7 8 9 10 11 12 13 14 15

ℵ6

The Student

Georg Ferdinand Ludwig Philipp Cantor was born in St. Petersburg, Russia, on March 3, 1845. To this day, his family's origins remain shrouded in mystery. Georg's father, Georg Woldemar Cantor, was born in Copenhagen. From the information on his Danish passport, he was born in 1809, but his gravestone in Heidelberg lists his birth year as 1814. We do know that the father's family migrated to St. Petersburg following the British siege of Copenhagen in 1807, and that the Cantors lost their home and possessions during the British bombardment. As a result, Georg Woldemar harbored anti-British feelings throughout his life.

Cantor's father was a devout Lutheran. The mother, Maria Bohm, was born a Roman Catholic. They were married in a Lutheran ceremony in St. Petersburg in 1842. We know, however, that the family had Jewish origins—most likely on both parents' sides, although certainly on the father's side, as the name "Cantor" suggests.[8] In a letter to a friend late in life, Cantor wrote that he had "Israelitisch" grandparents.

Two of the grandparents were Georg Woldemar's Danish parents: Jacob Cantor and his wife, whose maiden name was Meier. Both Cantor and Meier are common Jewish names. Quite possibly the grandparents on the Bohm side were Jewish as well.

Georg Cantor was the first of the Cantors's six children. Georg's younger brother, Louis, immigrated to the United States in 1863, and in a surviving letter he wrote that year to his mother from Chicago we find the sentence: " . . .we are the descendants of Jews." This supports the assertion by the historian of mathematics E. T. Bell that both sides of the family had Jewish roots.[9] The issue of whether or not Georg Cantor was Jewish—by origin, by belief, or by cultural values—plays an important role in our story.

In 1856, the Cantor family relocated to Frankfurt, Germany, following the father's contracting a pulmonary disease, which was exacerbated by the damp Baltic climate. Within a few years, however, Georg Woldemar died of consumption. Back in St. Petersburg, Woldemar owned a successful international wholesaling company, Cantor & Co., whose business interests reached from Europe to the United States and Brazil. By the time he retired in Germany, Woldemar had amassed a considerable fortune. While living in comfortable retirement in Frankfurt, Woldemar spent his time writing letters to his son Georg, who was away at the gymnasium and later living in Switzerland. These letters helped guide young Georg in setting the course of his career.

The Cantors had strong musical talents and family members played various instruments and taught music. Georg

Woldemar's cousin Joseph Grimm was a famous Russian chamber music player at the Russian Royal Court. On his mother Maria Bohm's side, Cantor was related to Joseph Bohm, the conductor and founder of the Vienna conservatory. Georg grew up with music and art, and a surviving drawing he made as a child shows considerable talent. One of Cantor's uncles was a professor of law at the University of Kazan, where he refined the legal machinery that later helped start the Russian Revolution. One of his students—as the family loved to point out—was Leo Tolstoy.

The young Cantor attended private schools in Frankfurt, and at age 15 was admitted to the Darmstadt Gymnasium. In a letter of those early days at the gymnasium, Georg Woldemar wrote his son:

"I close with these words: Your father, or rather your parents and all other members of the family both in Russia and in Germany and in Denmark have their eyes on *you* as the eldest, and expect you to be nothing *less* than a Theodor Schaeffer and, God willing, later perhaps a *shining star* on the horizon of science."[10]

Theodor Schaeffer was Cantor's teacher at the gymnasium, and apparently the father saw in him a model for his son's future success. Georg Cantor kept this letter from his school days by his side, as if to draw from his father's words the strength needed to fight his difficult way through life.

While still an adolescent, Cantor was drawn to mathematics. At the age of 15 he wanted to concentrate on mathematics and sought the approval of his father. In the spring of 1862, he wrote to his father, following a decision he had made:

"My Dear Papa!

You can imagine how very happy your letter made me; it determines my future. . . . My sense of duty and my own wishes fought continuously one against the other."[11]

Subsequent letters from his father indeed encouraged Georg to study mathematics as well as physics. In one letter he expressed the desire that his son study astronomy as well, and recounted a dream in which he himself was looking at the heavens through a telescope, marveling at the infinitude of stars. At any rate, Georg Woldemar clearly imparted to Georg a strong ambition to succeed in his academic endeavors. He also influenced him religiously. Some have argued (E. T. Bell among them) that the relationship between the autocratic father and the obliging son contributed to the younger man's mental problems later in life.

In August 1862, Cantor took his school examinations. He passed them with high marks and was now qualified to study sciences at the university. Cantor performed better in the exact sciences than in geography, history, and the humanities, and it was thus that the authorities decided he should concentrate on science.

Later that year, Cantor began to study mathematics at the Polytechnical Institute of Zurich. Soon, however, he was able to transfer to the more prestigious University of Berlin. Cantor's move to Berlin offered him a golden opportunity to learn mathematics from the world's masters. He took courses from Karl Weierstrass, Ernst Eduard Kummer, and Leopold Kronecker. Although he excelled in every subject he took at the university, Cantor was attracted to the theory of numbers. In 1867, he wrote a brilliant dissertation in this area.

Credit: Amir D. Aczel.

The main building at Halle University.

Cantor's dissertation was on a problem in number theory studied by Gauss. Thereafter, he continued to study the Gaussian theory of numbers and made important contributions to this subject, which were published in mathematics journals over the next few years.

Following the receipt of his doctoral degree, Cantor took the first position he was offered, that of a *Privatdozent* at the University of Halle. In this entry-level position at German universities, an instructor tutors students privately, living from whatever pay students provide. Cantor spent the rest of his time conducting intensive research in mathematical analysis, influenced by the ideas of Weierstrass. It was this kind of work that later brought him into direct conflict with his former Berlin professor, Leopold Kronecker, and resulted in a lifelong confrontation.

At Halle, Cantor began the study of functions based on Weierstrass's methods, leading him to the concept of convergence. He was deeply involved with the methods of potential infinity used in mathematics since the early Greeks, later refined and modernized by the analysts at Berlin.

$\aleph_7$

The Birth of Set Theory

At Halle, Cantor settled down to the mediocre life of an academic at a second-tier institution. Here, no great ideas were discussed at the mathematics department meetings, and there were no wonderful colloquia with great speakers on exciting new research topics.

By this time, Cantor had married Vally Guttmann, a friend of his sister, who came from a Berlin Jewish family. The two had met in Berlin and were married a few years after moving together to Halle, in 1875. They began to raise a family on Cantor's modest government income in the provinces, which was significantly less generous than the compensation at the University of Berlin. But it was here in a small city in the German countryside that Cantor developed a complete mathematical theory all his own.

Mathematical research is best done within a community of good mathematicians. Research results can be shared and ideas exchanged, so that new theories can develop and thrive. Working in isolation is hard and slow going, and there are many blind alleys into which a mathematician can stray when there is no possibility of sharing ideas with colleagues. But Georg Cantor was somehow able to produce one of the most amazing theories in the history of civilization—working alone.

Cantor brought with him to Halle some powerful, important ideas in mathematical analysis, which he learned in Berlin from Karl Weierstrass. At Weierstrass's course on function theory, one of the best mathematics courses ever given, Cantor had been exposed to unique concepts. Weierstrass expounded on the technique he had developed following Bolzano's ideas on limits and infinite sequences, and came up with an insightful definition of an old discovery: irrational numbers. The Bolzano-Weierstrass approach to irrational numbers was based on limits and on the property of spaces the two had independently discovered, which states that an infinite sequence in a bounded space contains a limit point in the space.

In the Bolzano-Weierstrass framework, built on the ideas of the ancient Greeks, we define an irrational number as a limit of rational numbers. The distance from the members of the sequence to the irrational number that acts as the limit point keeps getting smaller. This is a similar mechanism to the one intrinsic to Zeno's paradox about a person who can never leave a room. The person walks half the distance to the door, then half the remaining distance, and so on ad infinitum. Here, the door may be taken to represent some irrational number, a limit of an infinite sequence of rational-number steps.

While he was in Berlin, Cantor's work remained under the influence of the Weierstrass tradition. In Halle, he continued pursuing mathematical analysis along these same lines. Weierstrass, the old high school teacher whose brilliance brought him a professorship, did not believe in publishing his results and did not even like to have his students take lecture notes at his course. One reason his work survives is that a Swedish student of Weierstrass, who later became an important mathematician in his own right and a good friend to Cantor, took careful notes and organized them once he was back in Stockholm. This student was Gösta Mittag-Leffler (1846–1927).

There is a persistent rumor among mathematicians that Mittag-Leffler is the cause for there not being a Nobel Prize in mathematics. It is said that Alfred Nobel developed an intense dislike for the mathematician, and to prevent the possibility that Mittag-Leffler would ever receive a Nobel Prize for his work in mathematics, Nobel decided that there should be no such prize. Mathematicians have taken their collective punishment (if indeed that is what it was) in stride, and to this day, the highest prize in mathematics is the Fields Medal, awarded in recognition of outstanding achievement in mathematics.

By all accounts, Gösta Mittag-Leffler was not only an outstanding mathematician, but also a decent human being. During Cantor's darkest hours, when no one would listen to his fantastic ideas about infinity, let alone publish them, Mittag-Leffler regularly published Cantor's work in his journal, *Acta Mathematica*. He had married a fabulously wealthy woman, and used her family's fortune to support mathematics. In the 1880s, he built a sumptuous villa in a suburb of Stockholm, which he bequeathed to mathematicians as an institute of mathematics. (Perhaps Nobel felt that Mittag-Leffler had done enough for mathematicians so they didn't need a Nobel Prize.) Mittag-Leffler approached Georg Cantor early in his career, recognizing Cantor's great potential—something others had not fully appreciated—and had Cantor translate some of his early papers into French, bringing him recognition through their publication.

During his early years at Halle, Cantor had another good friend and benefactor, Richard Dedekind (1831–1916). Cantor needed *Acta Mathematica* as an outlet for his research activity because other doors had been closed to him since Kronecker and Kummer opposed his work. In Berlin, the only mathematician who supported Cantor's new work on infinity was Weierstrass. The admiration was mutual, and throughout his life, Cantor always spoke highly of Weierstrass and of his methods in mathematical analysis. A paper Cantor wrote in 1877 was almost rejected for publication in a journal edited by Kronecker. What saved the day for Cantor was Dedekind's intervention on his behalf.

Cantor first met Dedekind while vacationing in Switzerland in 1872. Dedekind was then a professor of mathematics at the Polytechnic Institute at Brunswick. The two became good friends. Dedekind was the last student of Gauss, and was born in Brunswick—Gauss's birthplace. As a young student, Richard Dedekind was interested in physics and chemistry, but when he enrolled at the University of Göttingen in 1850, he developed a love of mathematics. In 1852, at the age of 21, Dedekind received his Ph.D. in mathematics with a dissertation on integrals, written under Gauss's supervision. Dedekind taught at the Brunswick Polytechnic for fifty years, and apparently did not aspire to be promoted and move to a better school.

Dedekind's greatest contribution to mathematics was in the area of irrational numbers and their definition. Dedekind invented the concept of a *cut*. A cut on the number line separates all the rational numbers below a given number from all the rational numbers above it. If the cut itself does not define a rational number, then the number is irrational. For example, the square root of two, an irrational number, is defined by the cut that separates all the rational numbers whose squares are greater than two from all the rational numbers whose squares are less than two. Dedekind's work with irrational numbers, and hence with infinities, made him a natural ally of Cantor.

Dedekind lived so long that a newspaper article erroneously listed his death years before his actual passing. The obituary amused Dedekind so much that he wrote a letter to the paper describing how he had spent the day on which he supposedly passed away. "I passed this day in perfect health and enjoyed a very stimulating conversation with my honored friend Georg Cantor of Halle," he wrote.

What Dedekind did not tell the newspaper was that for the seventeen years prior to 1899, he had not heard a word from his "honored friend." The reasons for the break in communications were as complex as Cantor's personality. The two mathematicians were very close in the early days, and were bound together by personal empathy as well as mathematics. Both mathematicians were pioneers researching irrational numbers and infinity. Dedekind cuts were one approach to the problem of irrational numbers, and Cantor's wide array of methods of infinity were another. It was only natural that Georg Cantor, who felt lonely at Halle, was eager for a position in Berlin or another major center; or, failing that, that a good mathematician—such as his friend Dedekind—join the faculty at Halle.

When a position of professor at a German university became available, a rigid bureaucratic procedure was followed to fill it. The faculty members at the institution with the expected vacancy had to draw up a list of candidates for the new opening. The list was rank-ordered by the faculty and submitted to the German Ministry of Education. The ministry then considered the list, and, assuming it approved the order of the candidates, offered the position to the first person named. If that person accepted the position, the process was over. But if the person declined, the ministry moved to the next person on the list. If all named persons declined, the faculty at the institution was asked to draw up a new list.

Georg Cantor was the professor of mathematics at Halle, and once a vacancy for a second professorship materialized, he was responsible for drawing up the list of candidates. Cantor, with the agreement of others at Halle, placed Dedekind at the top of the list, and sent it off to the Education Ministry. Politely, Dedekind declined the offer. As for his reasons for declining, Dedekind listed financial considerations—the Brunswick position allowed him more financial freedom. Cantor was crushed by his friend's decision, and for the next seventeen years the two mathematicians did not exchange a single letter. Perhaps this was the first hint of the emotional instability that plagued Cantor in later years. In the meantime, Cantor founded the mathematical theory of sets.

Georg Cantor came to the idea of infinity—actual infinity rather than the potential infinity of limits mathematicians had used for centuries—not from considering numbers directly, but from considering *sets*. Cantor started by looking at numbers, points on the line, that *converged* to a limit point. A limit point of a set is a point that is arbitrarily close to members of the set. He came to this line of thinking from the Weierstrassian idea of defining irrational numbers as limits of rational sequences. Then Cantor decided to look at the *set* of limit points of a given set of points. For example, the set of irrational numbers in an interval is the set of limit points of rational numbers in the interval. He then asked himself the question: "What would happen if I now looked at the set of all limit points of the set of limit points?" And he continued his reasoning onwards in this way. Defining a set P, he called the set of its limit points the *derived set*, P'. Now the set of limit points of the limit points was denoted P''. From there he got P''', P'''', and then P''''', . . . , $P^{(infinity)}$,. . . . Cantor was interested in the question: "When do the derived sets, if ever, become empty?" That is, was there a case in which no more limit points could be extracted? But Cantor was intrigued by the mere process of continuing to construct the derived sets, starting with infinite sets and continuing through more infinite sets. Ultimately, this conundrum led him to the key issues of his research—his life's work on the nature of infinity itself. Cantor's monumental work on infinity began with an investigation of the nature of sets. In addition to being the first person in history to deal with actual infinity, Cantor also became known as the father of the theory of sets.

To be sure, the elementary theory of sets had its origins long before Cantor. Any time we classify things we use set theory, in a rudimentary way. In his classic book *Naïve Set Theory*, Paul R. Halmos wrote: "A pack of wolves, a bunch of grapes, or a flock of pigeons are all examples of sets of things. The mathematical concept of a set can be used as the foundation for all known mathematics."[12] And in fact, what we call today the *foundations* of mathematics is an area that includes set theory and logic. From things such as grapes or wolves—elements in a collection—the entire edifice of modern mathematics can be constructed. The great Italian mathematician Giuseppe Peano (1858–1932) defined the concept of a *number* using set theory in an ingenious way, as we will soon see.

Starting from the concept of a set, a collection of items or people, other sets can be constructed by using set operations. These operations correspond to the words *and*, *or*, and *not*. The union of two sets is the set whose members belong to one set *or* the other (or both). The intersection of two sets is the set of all elements that are members of one set *and* the other. And the complement of a set consists of all points that do *not* belong to the original set. Using the *or, and*, and *not* operations, we can define an interesting rule for sets that is useful in computer science (in the context of yes-no Boolean operations). The rule is due to Augustus De Morgan (1806–1871), who was one of the founders of the British Association for the Advancement of Science. The rule says:

$$\text{Not}(A \text{ or } B) = (\text{not } A) \text{ and } (\text{not } B)$$

This is demonstrated in the picture below. Look at the figure and convince yourself that the area that is outside the union of *A*, *B* is indeed the area that is *both* outside *A* and outside *B*.

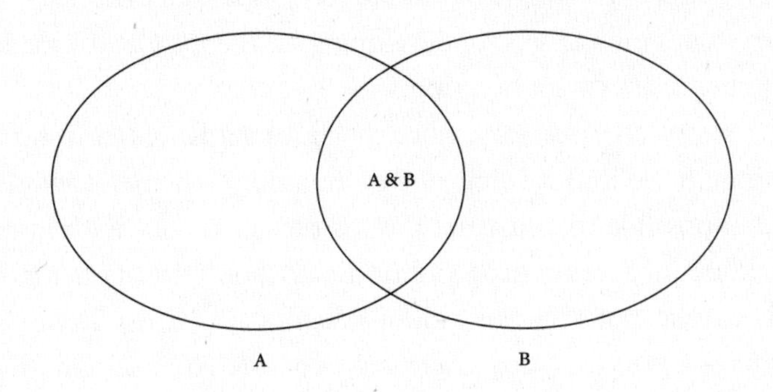

One of the key elements of set theory is the famous *empty set* or *null set*: the set containing no elements at all. The empty set is everywhere—it is a subset of every set. Why? By contradiction: for the statement not to be true, we would have to exhibit a point that belongs to the empty set but does not belong to a given set, *A*. But since the empty set has no elements, no such element can be exhibited, so the statement is true.

Set theory leads invariably to great paradoxes. Since the foundations of mathematics consist of the theory of sets along with elements of mathematical logic, the paradoxes of set theory make the entire foundations of mathematics problematic. We may find it hard to believe that an elegant and seemingly very simple system of numbers and operations such as addition and multiplication—elements so intuitive that children learn them in school—should be fraught with holes and logical hurdles. But such is mathematics. And when infinity is added to the mixture, the pitfalls multiply.

When Georg Cantor started developing his own set theory, he extended the results to infinite sets. Today, all of set theory deals with infinite sets. In developing his theory, Cantor made implicit use of a set of axioms.

Axioms are necessary, since they give mathematics a starting point. Without some sort of an axiom scheme, it would be impossible to construct any consistent theory with logic guiding its development and results. Among the mathematicians who put forward axiom schemes was Ernst Zermelo (1871–1956). In 1904, Zermelo constructed an alternative axiomatic system for set theory. This system later became known as the Zermelo-Fraenkel set theory (ZF), adding the name of the logician Abraham Fraenkel (1891–1965). But the ZF system, like its predecessors, was not to be free of paradoxes.

Even before Cantor weighed in with the overwhelming power of actual infinity, the system was plagued with inconsistencies. Sets were contemplated that were simply too huge, too incomprehensible to the limited human mind, to be amenable to taming by axiomatization.

But people tried nonetheless. And while the serious paradoxes of set theory would continue to beset humanity to our own time, the theory of sets was still useful in defining the concept of a number, as was brilliantly done by Giuseppe Peano. Starting in the 1880s, Peano began to study how numbers could be defined from the concept of a set using only set operations. He was helped in his work by a school of mathematics he had founded at the University of Turin, including many gifted mathematicians such as Cesare Burali-Forti (1861–1931) and others.

Peano's foundations underlie all of mathematics. From them we can define not only the natural numbers but also the rational numbers, the real numbers, including irrational ones, complex numbers, and all of arithmetic. Peano's elegant derivation of a number system from sets is demonstrated below.

Peano defined zero as the empty set. One was then defined as the set containing the empty set. Two was the set that contained the empty set and the set containing the empty set. The process was then assumed to continue ad infinitum, defining every whole number. In terms of notation, the Peano number system is as follows:

$$\varnothing \ \{\varnothing\} \ \{\varnothing, \{\varnothing\}\} \ \{\varnothing, \{\varnothing\}, \{\varnothing, \{\varnothing\}\}\}$$

(for the numbers 0, 1, 2, and 3. The rest follow by simple extension of the same principle).

So while paradoxes plagued and continue to plague the theory of sets, the discipline survives and continues to form the foundation for the entire field of mathematics. Cantor's genius was that he was able to arrive at actual infinity, and learn important truths about it, starting with sets. The sets he first defined were the powers, *P*, beginning with a set and then deriving the set of its limit points. Here, a set must necessarily be infinite. An infinite sequence in a bounded space, as Bolzano and Weierstrass have taught us, will have a limit point. When there are many sequences converging to many points, we may define the *set* of limit points of the original set. Then Cantor considered the limit points of this new set, and so on. Soon enough, he had in his hand an infinite collection of sets with infinitely many points.

At this point, Cantor was about to outdo Peano. In the next stage of his research, Cantor defined a whole new world of numbers. These were not ordinary numbers (numbers we now call *ordinals*), as Peano had defined them. All of Cantor's numbers were beyond the finite world. They were called *transfinite*. These numbers were extensions of the concept of a number to the unknown, mysterious world of actual infinity. But among other perils on his road to the secret garden, Cantor would encounter a formidable principle with bizarre, disturbing consequences: the axiom of choice.

$\aleph_8$

The First Circle

The famous French mathematician Henri Poincaré (1854-1912) said that Cantor's set theory was a malady, a perverse illness from which some day mathematics would be cured. In response, however, the eminent German mathematician David Hilbert said that "no one would expel us from the paradise that Georg Cantor has opened for us." Cantor's entrance into the infinite Garden of Eden indeed opened a new era in mathematics. The mysterious world of actual infinity can be visualized, as Dante would have described it, by imagining circles nested within circles. Each circle signifies a loftier position—a higher order of infinity. The lowest form of all of these infinities is the level of infinity occupied by the natural numbers 1, 2, 3, etc.

The natural numbers can be counted, even though they are infinite. It is the process of counting that matters, not the actual counting, since such counting will never end. The natural numbers *can* be counted, since it is possible to call them one after the other, 1, 2, 3, 4, . . . and so on. Thus—while infinite—the natural numbers are *countable*. Early on in his career, Cantor used an ingenious argument to show that the *rational numbers* are countable as well. Thus, as Galileo had shown that there are as many squares of whole numbers as there are whole numbers, Cantor showed that there are as many rational numbers as there are whole numbers. The argument is called Cantor's Diagonalization Proof of the Denumerability of the Rational Numbers.

Cantor first used the proof in 1874, but later, in 1891, he improved the proof because he was worried about a number of technical implications, which were not clear to him back in 1874. In 1891, he thought the proof could become powerful enough to allow him to establish an entire hierarchy of transfinite numbers. Cantor began his proof by arranging all the rational numbers in a two-dimensional array, as shown below.

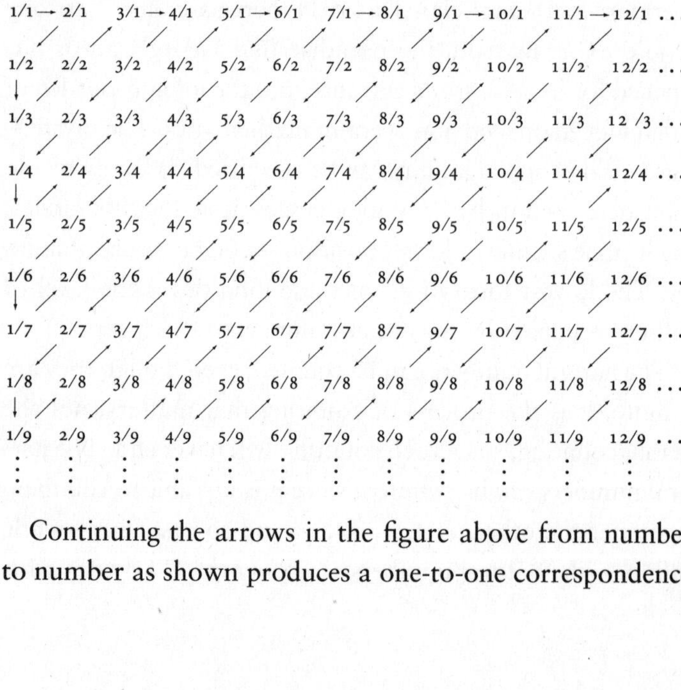

1/1 → 2/1 3/1 → 4/1 5/1 → 6/1 7/1 → 8/1 9/1 → 10/1 11/1 → 12/1 ...

1/2 2/2 3/2 4/2 5/2 6/2 7/2 8/2 9/2 10/2 11/2 12/2 ...

1/3 2/3 3/3 4/3 5/3 6/3 7/3 8/3 9/3 10/3 11/3 12 /3 ...

1/4 2/4 3/4 4/4 5/4 6/4 7/4 8/4 9/4 10/4 11/4 12/4 ...

1/5 2/5 3/5 4/5 5/5 6/5 7/5 8/5 9/5 10/5 11/5 12/5 ...

1/6 2/6 3/6 4/6 5/6 6/6 7/6 8/6 9/6 10/6 11/6 12/6 ...

1/7 2/7 3/7 4/7 5/7 6/7 7/7 8/7 9/7 10/7 11/7 12/7 ...

1/8 2/8 3/8 4/8 5/8 6/8 7/8 8/8 9/8 10/8 11/8 12/8 ...

1/9 2/9 3/9 4/9 5/9 6/9 7/9 8/9 9/9 10/9 11/9 12/9 ...

⋮ ⋮ ⋮ ⋮ ⋮ ⋮ ⋮ ⋮ ⋮ ⋮ ⋮ ⋮

Continuing the arrows in the figure above from number to number as shown produces a one-to-one correspondence of the rational numbers with all the natural numbers. Here 1/1 is paired with 1; 2/1 is paired with 2; 1/2 is paired with 3; 1/3 is paired off with 4; and so on.

Thus every rational number is counted off against a natural number (even though there are redundancies; for example, the number 1, appearing in the guise of 2/2, 3/3, etc., is counted infinitely many times).

The process yields a very surprising result. While the natural numbers, or the integers—now including zero and all negative integers—are separated from each other by one unit, the rational numbers seem much more numerous because we know that they are packed more densely than the integers. The rational numbers are mathematically *dense* in the set of real numbers, meaning that in any infinitesimally small neighborhood of any number on the real number line we can find rational numbers. And yet the proof is solid and there is no mistaking the property: there are as many rational numbers as there are integers. The order of infinity of the integers and the order of infinity of the rational numbers is the same. This is the first circle of the secret garden of infinity.

Now we may wonder whether every infinite set can be enumerated against the integers. All we'd have to do is pair off the infinite set of numbers against the integers and we would then have a one-to-one correspondence with the integers, showing that there are as many (infinitely many) numbers in the set as there are integers. But in 1874 Cantor proved that there are circles of infinity that lie beyond the first circle. He proved that the set of irrational numbers *cannot* be enumerated. It is not possible to find a one-to-one correspondence between all the real numbers on the line (irrationals and rationals put together) and the integers. Cantor proved this striking result by an interesting method.

Cantor and his friend Dedekind had been advocating theories about the real numbers that hinged on the assumption that there was something much deeper about continuity than simply the stringing together of infinitely many numbers. Indeed, the irrational numbers are infinitely richer than the rational and the algebraic numbers (irrational numbers that are roots of equations with rational coefficients, and thus also countable as the rational numbers). This infinitely-richer structure of the transcendental irrational numbers such as π or e (the base of natural logarithms) fills the gaps between all the rational and algebraic numbers and provides the real line with its continuity. Cantor had posed a question about the transcendental numbers to Dedekind in a letter of 1872, and spent the following year studying the problem. It was just before Christmas of 1873 that he obtained his ingenious proof that the transcendental numbers (or the real numbers containing them, to be more general) were of such a high order of infinity that they could not be counted.

Cantor started by assuming that—just as he had done with the rational numbers in his diagonalization proof—there *is* some way to enumerate all the numbers on the real line. Cantor began by restricting his analysis to the numbers between zero and one. Cantor then assumed that the real numbers from zero to one could be listed in order. Then he tried to match each one of these numbers with one integer. The first number on the list can be matched with the number 1, the next one with the number two, and so on. He assumed that all the numbers between zero and one can thus be listed (here shown in no particular order):

0.12421567437895 43 . . .
0.23411762998295 47 . . .
0.77639823965466 11 . . .
0.48295344790123 75 . . .
0.03481094321629 84 . . .
. . . .
. . . .

Cantor took the above as an infinite list of the infinitely many numbers between zero and one. But at that moment, Cantor made a stunning realization. He noticed that he was able to define a diagonal number, which he constructed by taking a digit from each of the infinitely many numbers above, using the first decimal from the first number on the list, the second from the second number, and so on to infinity. This number is: 0.13691. . . .

Cantor now used a clever device. He changed each one of the digits of the diagonal number. This transformation can be achieved, for example, by adding 1 to each of the digits, giving him the new number: 0.24702. . . . Now, this new number is different from all (infinitely many) numbers on the list above, because it is different from every single number on the list at least in its value for the particular digit taken from that number (since one was added to that digit). Since the new number Cantor created was different from all the numbers on the list, it is impossible to list all the numbers between 0 and 1. This proved that the size of the set of all real numbers is greater than the (infinite) size of the set of all integers and all rational numbers.[13] How much bigger is the set of all real numbers from the set of integers Cantor could not say.

It should be intuitive that the listing of all the real numbers is impossible. We know that for any number on the line there is no "next" number. The numbers on the line are infinitely dense.

Cantor showed with the proof above that there are different orders of infinity. There is the order of infinity of the rational numbers, and there is another order of infinity that characterizes all the real numbers on the number line. But, while Cantor knew that the order of the real numbers was higher, he could not say whether it was the "next higher" infinity after that of the rational numbers, or whether there was some intervening order of infinity. Next, Cantor would address the question of dimension.

Before doing so, Cantor decided that he had to try to publish his important result. Cantor knew that in Berlin there was strong opposition to his work on the irrational numbers and the sizes of sets. He decided, therefore, to publish his results hidden within a paper whose title would not hint at its controversial contents. He knew that before the paper's publication, a number of mathematicians might peruse the title of the work to see if there was anything objectionable. In such a case, they would try to convince the editor of the journal not to publish it. In particular, Cantor was leery of Leopold Kronecker, who had already indicated his reservations about the validity of Cantor's work.

Cantor named his paper "On a Property of the Collection of All Real Algebraic Numbers." Of course the paper was about the property of the *remaining* numbers on the real line, once the countable collection of rational and algebraic numbers had been removed. The main result was that these remaining numbers, the transcendental irrational numbers, could not be enumerated—their order of infinity was higher than that of the rational and algebraic numbers. The ploy worked, and the paper was published in *Crelle's Journal* later that year.

ℵ9

"I See It, but I Don't Believe It"

On June 29, 1877, Cantor wrote a letter to Dedekind. He was very excited, and totally bewildered. He could prove mathematically the property of infinity he had just discovered, but the result made no sense at all to him. In his letter, he wrote—uncharacteristically—in French: "Je le vois, mais je ne le crois pas." Cantor had just discovered a property of infinity that even he found shocking.

The idea of dimension is crucial to all of mathematics. Euclid defined a point as something that had no length; a line as something that had no breadth; and a plane as something that had no depth. A line has length, a plane has area, and a three-dimensional object has volume. Continuing to higher dimensions is very natural in mathematics, although our three-dimensional intuition does not continue further.

A POINT (DIMENSION 0) A LINE (DIMENSION 1) A PLANE (DIMENSION 2)

Cantor asked himself the question: What is the order of infinity of various objects of different dimension? To answer this question, Cantor appealed to the work of the great French mathematician and philosopher René Descartes (1596–1650). Descartes was the best-known mathematician of his time, an era that saw a number of great mathematicians such as Pascal, Fermat, and Galileo.

René Descartes was born on March 31, 1596, in La Haye, near Tours, in France. His family belonged to the nobility, but was not wealthy. He was the third child, and after his birth his mother died. The father remarried, and René and his siblings were raised by a governess. As a child, he became known as a young philosopher because of his great curiosity about the world, always wanting to know why things were the way they were. At the age of eight, René was sent to be educated at the Jesuit College at La Flèche.

Descartes was a frail boy, and his health was poor. At the college, the rector wanted the boy's health to improve, so he allowed René to sleep late in the mornings and stay in bed until he decided that he felt well enough to join the others in the classroom. This privilege started Descartes on a lifelong habit of staying in bed and considering mathematical and philosophical problems while at rest. Later in life, Descartes would admit that the foundations of his philosophy and mathematics were developed during the time he spent in bed at the Jesuit College. Even though at the college he studied Latin and Greek and rhetoric, his mind always wandered to philosophical questions and mathematical problems.

Upon graduation, Descartes entered the University of

Poitiers to study law. He soon discovered that he was not interested in the law, but instead wanted to see the world. First he moved to Paris and spent his time gambling successfully and earning significant amounts of money. Then he traveled to Holland and trained to become a soldier. He took part in a number of military campaigns in Holland, and then joined the Bavarian army, fighting against the forces of Bohemia. The army was wintering on the banks of the Danube, and Descartes spent his time in bed. During the night of November 10, 1619, Descartes experienced three very vivid dreams.

In the first dream, Descartes saw himself blown by winds from the safety of a shelter in his church toward another place. There, the wind could touch him no more. In the second dream, he saw himself observing a violent storm, but through the eyes of a scientist. This allowed him to escape the storm's power since he could understand its nature and disarm the storm in some sense and remain unaffected by its physical violence. In the third dream, Descartes was reciting a poem of Ausonius, beginning with the words; "What way of life shall I follow?" Waking up, Descartes was filled with new enthusiasm and a sense of mystery. He concluded that the dreams had given him a magic key by which he could unlock the secrets of nature and tame its powers.

Descartes' dreams of obtaining the key to the secrets of nature became a reality when he developed the field of analytic geometry. Descartes was able to apply algebra to geometry, and thus to find a way of assigning numbers to elements of geometrical figures. Descartes discovered the coordinate system now named after him, the *Cartesian coordinate system*. But the philosopher-soldier took another eighteen years to publish his breakthrough ideas, and in the meantime continued on the campaign trail of various European conflicts.

In the spring of 1620, Descartes saw heavy fighting in the battle of Prague and almost died. Throughout this period, Descartes had an internal conflict as well. He was torn between religious feelings and the call of science, and saw a possible contradiction between the two areas. After a stint fighting for the Duke of Savoy, Descartes settled for three years of meditation in Paris. Following the example of Galileo, he spent much time during these quiet years looking at the stars through a telescope. But he made no astronomical discoveries and probably gazed at the stars for meditative reasons.

A Catholic clergyman convinced Descartes that it was his duty to God to publish his scientific and philosophical discoveries. At the age of thirty-two, Descartes became convinced that it was indeed his religious duty to publish, and in order to do so he moved to Holland, where the printing presses were most productive during that period. Descartes spent the next twenty years wandering throughout Holland and carrying out an active correspondence with European intellectuals. His closest friend was Father Marin Mersenne (1588–1648), a French clergyman Descartes had known since his early days at the Jesuit College at La Flèche. Mersenne was a mathematician and the discoverer of the now-famous Mersenne prime numbers. The two men discussed mathematical and philosophical problems.

Descartes is now known as the "father of modern philosophy" thanks to the publication of his book *Discourse on the Method of Reasoning Well and Seeking Truth in the Sciences* in 1637. Descartes' ideas of analytical geometry were written in his important book, *La Geométrie*, which he published as an appendix to *Discourse on the Method*. Another of his books, *Le Monde*, was dedicated to his friend Mersenne. This book was his attempt at a scientific description of the creation of the world—a revision of the Book of Genesis in an attempt to reconcile science with religious belief. Before the book was published, Descartes received news of Galileo's trial by the Inquisition. Galileo's book was much milder than *Le Monde*. Thus, fearing a fate as unpleasant as Galileo's, or worse, Descartes chose to withhold his book from publication. The book was published only after his death.

Ironically, Cardinal Richelieu gave Descartes an unprecedented license to publish anything he wanted to publish in France or abroad. In Holland, however, Protestant theologians condemned his writings as atheistic. Eventually, his works were listed on the Church's Index of Prohibited Books.

In 1646, Queen Christine of Sweden invited Descartes to join her court as the Royal Philosopher. Descartes liked his peaceful life in Holland and was reluctant to move to Sweden. But the queen persisted, preying on his admiration and respect for royalty. Finally, in the spring of 1649, the queen sent Admiral Fleming of the Swedish Royal Fleet on a special mission to pick up the philosopher and bring him to Sweden.

The French ambassador to Sweden offered Descartes an

apartment inside the embassy complex in Stockholm, and Descartes accepted and lived there as a special guest of the ambassador. Early every morning, a royal carriage collected the philosopher, who until then had followed his usual practice of sleeping late. The queen had Descartes give her daily lessons in philosophy at 5 A.M. in the unheated library of her palace. His new Spartan lifestyle quickly took its toll on the 54-year-old philosopher, and early in 1650 he fell ill. He died on February 11. Two decades later, the French Government brought his remains back to France and reburied them in the Pantheon in Paris, among France's most honored dead.

Descartes' discovery of analytic geometry was an extension of the work of the ancient Greeks. They knew that there was a connection between numbers on the line (or plane or a higher-dimensional surface) and numbers. But the Greeks were never able to explore this relationship. Descartes' genius extended this knowledge. The Cartesian coordinates he discovered give us the method to study the numerical properties of curves and functions in a given space. Descartes divided the plane into four quadrants. These quadrants meet at a point called the origin of the coordinate system. This partition gives us the now familiar *X-Y* plane. At the origin, the values of both *X* and *Y* are zero. Moving right increases the value of X, and moving left decreases it. Moving up increases the value of *Y* and moving down decreases it. Every point on the (infinitely large) plane has an *X*-coordinate and a *Y*-coordinate. This feature, the quantification of every point on the plane, is Descartes' greatest contribution to mathematics and the applied sciences.

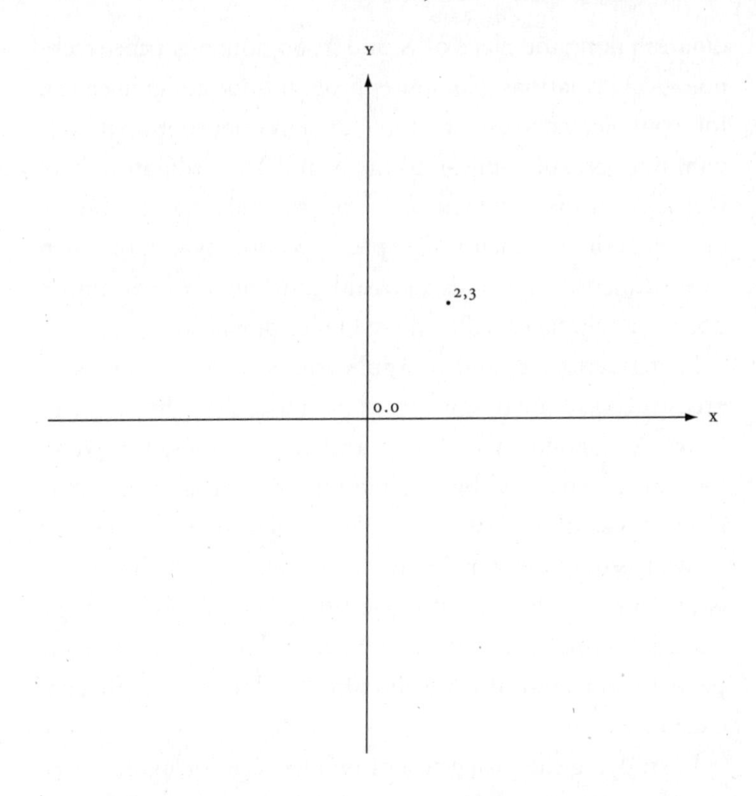

The Cartesian coordinate system is so useful in science that we are not even aware of the many important applications made of it daily. The pixels on a computer screen or television screen are all digitized according to the Cartesian coordinate system. As electrons stream continuously through a wire, they are projected onto the two-dimensional screen in such a way that every electron has an exact position to which it is sent, given by its specified *X*- and *Y*-coordinates. Maps work the same way, with East-West and North-South coordinates taking the place of *X* and *Y*. So do many other technological inventions. Computer programs for designing a car, for example, make use of a three-dimensional coordinate system. But here, in addition to the *X* and *Y* coordinates, there is a *Z*-coordinate, which measures the depth of a point in our usual three-dimensional space. Extensions to more than three dimensions are also possible, although our intuition does not help us visualize these higher dimensions.

In mathematics, and in applications of mathematics to areas such as statistics, we often use more than three dimensions. A respondent's answers to five questions, for example, can be analyzed by considering how close the answers are to those of another respondent when these answers are viewed as points in a five-dimensional space (each dimension is identified with a specific question). Such analyses, while not easily visualized because of the high dimensionality, make perfect mathematical sense and often lead to meaningful conclusions.

In studying the properties of infinity, Cantor made use of the Cartesian coordinate system. He asked himself the natural question that followed from his previous research: Are there more points on the plane than on the line? And, in general, how does the dimension of a mathematical entity determine the number of points it contains? As he had done in his previous studies, Cantor chose to look at the numbers between 0 and 1, knowing that there would be no loss of generality and that the results should apply to the entire number line. He knew that Bolzano had shown that any interval of numbers had the same number of points as any other line segment, so looking at the interval from 0 to 1 simplifies the analysis without sacrificing the generality of the conclusion. To start, Cantor drew a 1-by-1 square next to the closed interval of numbers between 0 and 1.

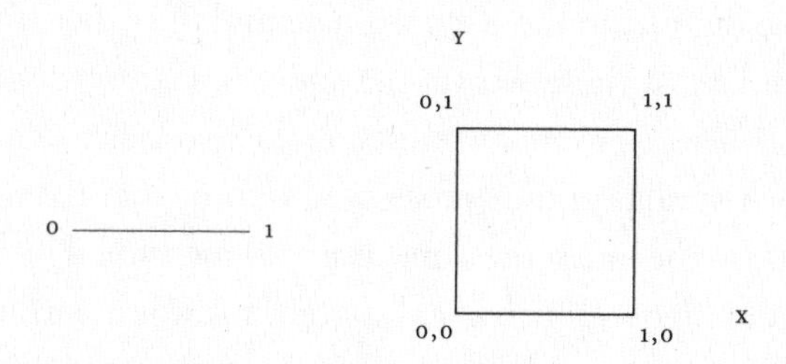

Following Descartes's ideas, Cantor knew that every point on the unit square could be represented uniquely by two numbers: its *X* coordinate and its *Y* coordinate. Every point on the 0 to 1 interval along the *X* axis was represented by a single number. All numbers, whether pairs of coordinates for a point on the square or a single number representing a point on the line, had the form: 0.23416573498451. . ., in general notation written as $0.a_1b_1a_2b_2a_3b_3$ . . . , because all of them were numbers between zero and one. Cantor wanted to test whether he could somehow construct a one-to-one transformation of the points on the square to the points on the line. To his great surprise (since he had believed there were more points on the square than on the line), he was able to do this.

Every point on the square is represented by a Cartesian *pair* of numbers between 0 and 1 (the point's *X* and *Y* coordinates): $0.a_1a_2a_3$ . . . , $0.b_1b_2b_3$. . . . Now, Cantor defined the transformation from the square to the line segment as follows: The transformation *alternates* the decimal expansion of the two coordinates, yielding the number: $0.a_1b_1a_2b_2a_3b_3$. . . .

This creates a unique number between 0 and 1. Thus, every point on the square (given by a pair of numbers) has a mate on the line segment (given by a point whose decimals alternate, one digit from the *X*-coordinate and the other from the *Y*-coordinate of the point on the square). Cantor concluded that there are just as many points on the plane as there are on the line. A similar argument showed Cantor that the number of points on the line is the same as in 3-dimensional, 4-dimensional, and higher-dimensional space. This was a baffling and completely unexpected finding. As far as infinity goes, dimension does not matter. Any continuous space, whether a line or a plane or an *n*-dimensional space, has as many points as the continuum. All these spaces are uncountable, as Cantor had proved earlier.

The puzzle of dimension was first raised in a letter Cantor wrote to Dedekind on January 5, 1874. He asked: "Is it possible to map uniquely a surface (suppose a square including its boundaries) onto a line (suppose a straight line including its endpoints), so that to each point of the line one point of the surface corresponds?"[14] Cantor went on to say that he thought the answer should be "no," since a surface has a higher dimension than a line, but that it was worth trying to establish such a correspondence to verify the impossibility of constructing the required transformation. Later that spring, Cantor visited Berlin and asked a number of his acquaintances what they thought of such an investigation. All of them thought that it was absurd to try to find such a correspondence since clearly several variables couldn't possibly be reduced to one variable.

Three years after his successful construction of what he thought was an absurd correspondence between lines and higher-dimensional surfaces, Cantor again wrote to his friend Dedekind, announcing his success. Continuous spaces of *n* dimensions, he found, had the same number of points (the same "power"—this is how mathematicians describe the existence of a one-to-one correspondence between sets) as the line. He wrote that he was well aware that this was a very controversial finding since mathematicians would find it hard to believe, and that it would upset many notions of geometry. He could *see* it, Cantor said, but he couldn't believe it.

Dedekind was very cautious in his response. He was a realist, and knew that the opposition to such revolutionary ideas would be swift and truculent. Dedekind congratulated Cantor on his proof, but warned him not to challenge traditional mathematical thinking as strongly as he had done in his letter to him. He wrote: "I hope I have expressed myself clearly enough. The point of my writing is only to request that you not polemicize openly against universal beliefs of dimension theory without giving my objections thorough consideration."

$\aleph_{10}$

Virulent Opposition

Until 1871, Leopold Kronecker was favorably disposed to the work of his bright former student at the University of Berlin, and offered Cantor help in establishing himself at Halle. Kronecker gave Cantor suggestions on his first papers dealing with trigonometric series and what would later be called the Cantor-Lebesgue theorem.[15] Cantor was impressed with the cleverness of a technique offered by Kronecker and was sincerely grateful to his former teacher for all his help.

But once Cantor began to extend his results and turn his attention to irrational numbers and infinity, Kronecker became increasingly uneasy. The problem started as a purely philosophical difference between two people. Kronecker had always opposed the ideas of mathematical analysis, and had continuing arguments with the father of analysis, Karl Weierstrass. According to tradition, Kronecker simply refused to believe that irrational numbers existed. And he wasn't impressed by that fact that circles inevitably lead to the number π. To Kronecker, only the integers were real. Everything else was pure fiction.

Since any child today can press the square root key on a calculator and obtain the square root of two (the unending decimals of this irrational number truncated after a finite number of places and rounded), we might think that Kronecker—who didn't believe such numbers existed—was a very poor mathematician. But Kronecker was actually a fine mathematician, and is well-known today for a number of important results in mathematics. The fact that he was a great mathematician made Kronecker's conflict with Cantor so much more dramatic. And their differences were deep—they were those of belief. Kronecker believed with his whole being that everything other than the integers was unnatural. Dealing with irrational numbers was an act against nature. And, in fact, he accused Cantor of being a "corrupter of youth" because he taught such concepts.[16] Along with his contempt for irrational numbers, Kronecker also harbored a deep hatred for anything even vaguely related to the concept of infinity.

Georg Cantor, on the other hand, deeply believed that infinity was God-given. To Cantor, infinity was the realm of God, and it consisted of various levels—the transfinite numbers. Beyond the transfinite numbers, there was an unreachable, ultimate level of infinity, the Absolute. The Absolute was God Himself.[17] On the lowest transfinite level was the infinity of the integers and the rational and algebraic numbers. The transcendental numbers and the continuous real line belonged to a higher level of infinity. As Cantor continued to explore infinity and the continuum, his war with Kronecker became more violent, and more personal.

The two men, the professor at Berlin and his former student now at Halle, did make an attempt to reconcile their differences. Cantor liked to spend his vacations in Germany's Harz Mountains, a small mountain chain an hour west of Halle. The tallest peak in these mountains is 3,000 feet high, and there are large wooded areas, meadows, and streams. Cantor occasionally met other mathematicians in the little villages and resorts of the Harz Mountains, where they would spend their time together discussing mathematical ideas in the restful setting of the forests. During one of his stays, Cantor summoned all his courage and wrote a conciliatory letter to his former professor, asking Kronecker to come and meet him at his mountain resort. To his surprise, Kronecker agreed.

The two men met in the Harz Mountains and discussed mathematics, and Cantor tried to explain to Kronecker his new discoveries and theories of the infinite. But ultimately the two mathematicians failed to reach an understanding. The gulf between their two philosophies was simply too deep to bridge. Kronecker could not accept the idea that irrational numbers existed and could not accept the veracity of analytical properties of the continuum. Soon after the two men parted, amicably on the surface, their enmity became even more bitter.

As soon as Cantor tried to publish his paper showing that dimension did not matter and that all continuous spaces, lines, planes, or higher surfaces, had the same order of infinity, Kronecker actively sought to prevent the paper's publication. Kronecker had opposed many results in mathematical analysis including the Bolzano-Weierstrass theorem, and had tried to dissuade other mathematicians from publishing results that used these theorems or dealt with irrational numbers and infinity.

Cantor sent his paper on the irrelevance of dimension to *Crelle's Journal* on July 12, 1877. The editor promised to publish the paper, and in Berlin Karl Weierstrass promised to promote its appearance in the journal. But Cantor was not sent proofs, and there was no indication that the paper was approaching publication. Cantor immediately suspected that Kronecker was acting behind the scenes to prevent publication. In anger, he complained to Dedekind and asked what Dedekind would suggest. Should he withdraw the paper altogether and try to publish it elsewhere?

Dedekind replied that he should wait and not get too excited about Kronecker. As it turned out, however, Kronecker was indeed the culprit, doing all he could to delay the publication of the paper. Kronecker argued with the editors that Cantor was dealing with empty, purely fictitious concepts. The transcendental numbers did not exist, he argued, and the mathematical public should be spared a meaningless paper. The editor, however, did publish Cantor's paper the following year. While Cantor was victorious in the end, he was shocked to find out how low his enemy was willing to stoop in his efforts to thwart him. Cantor never again sent a paper to the respected *Crelle's Journal.* His future work would be published elsewhere, away from Kronecker's web of deceit.

Cantor advocated, and followed in his research, a doctrine of open exploration of new concepts and ideas. He believed in freedom in mathematics and philosophy, allowing anyone to pursue ideas wherever these ideas might lead. On the other hand, the mathematics of Kronecker and his followers was safe but benighted. These conservative thinkers tried to base all of mathematics on concepts such as whole numbers and finite realms. They sought to restrict novel ideas, since these new concepts endangered their view of mathematics. Cantor believed that their arbitrary restrictions impeded the growth of the field, and in letters to other mathematicians he gave many examples of mathematical ideas that never would have had a chance to succeed and evolve if their proponents had encountered the kind of opposition he now faced.

But Kronecker and his allies in Berlin were unmoved. Weierstrass was a highly respected veteran professor. Whatever these purists might have thought about mathematical analysis, they would not have succeeded in battling such an established giant. Cantor was in a different position. A few years earlier, he had been a student, and Kronecker and his colleagues were his teachers. From their point of view, he should have been shaped in their image, and had no right to challenge his teachers' philosophy.

Cantor did not make things easy for himself, either. Not only did he have the brazenness to expect the world of mathematics to embrace his ideas of infinity, but he also believed that he belonged in the Berlin group of thinkers. He was embarrassed to be associated with a second-rate school, and believed that he was the only mathematician of his time to truly understand all of mathematics. He expected to be called any day to take up a professorship at the University of Berlin. But Kronecker knew Cantor's faults. He sensed that if he mounted a strong opposition—and made the attack personal—eventually Cantor would crack.

In September 1883, Cantor wrote letters to Mittag-Leffler and to the French mathematician Charles Hermite (1822-1901), bitterly complaining that he was being buffeted by Kronecker's campaigns against him, which by now had taken a personal turn. Kronecker was vilifying Cantor, calling him a charlatan and "a corrupter of youth," and referring to his work as "humbug." Cantor was besieged, lonely, angry, and frustrated. In the backwaters of Halle, far from the centers of mathematical activity, he could not even fight back effectively. In December, after further assaults on his work, Cantor became more enraged. He sought to retaliate against Kronecker, and in despair came up with a bizarre plan. He was now convinced that he could never obtain a professorship in Berlin since Kronecker, entrenched and powerful, would always stand in his way. So Cantor decided to apply for a professorship anyway—for the mere purpose of annoying his enemy.

By the end of the year, Cantor wrote Mittag-Leffler of his ploy and its results. "I knew precisely the immediate effect this would have," he wrote, "that in fact Kronecker would flare up as if stung by a scorpion, and with his reserve troops would strike up such a howl that Berlin would think it had been transported to the sandy deserts of Africa, with its lions, tigers, and hyenas. It seems that I have actually achieved this goal!"[18]

But it was now Kronecker's turn to strike back at Cantor. Kronecker wrote to Mittag-Leffler asking to publish in his journal, *Acta Mathematica*. Years earlier, Kronecker's actions had successfully driven Cantor out of *Crelle's Journal*. Now Kronecker was shrewdly trying to push Cantor out of the only mathematical journal that had a sympathetic editor interested in his work. Cantor suspected that Kronecker's paper would constitute an attack on his own work published in *Acta Mathematica*, the journal he considered his home turf, and would discredit him there, where it would hurt him the most. In frustration and fear, Cantor wrote to his friend Mittag-Leffler threatening to stop sending him his work. As it turned out, Kronecker had no paper to send to *Acta Mathematica*. Kronecker had simply pretended to want to publish in the journal in order to upset Cantor. The move succeeded. Cantor's response eroded his relationship with one of his few remaining friends, Mittag-Leffler.

The strain of these battles, which Cantor never stood a chance of winning, was taking its toll on his health. In May 1884, Cantor had his first nervous breakdown, lasting over a month. When the attack took place, Cantor was working very hard on an impossible mathematical problem. The frustration he felt at not being able to solve the problem, added to his other woes, was almost certainly a major cause of his illness.

$\aleph_{11}$

The Transfinite Numbers

A googol, according to *The Penguin Dictionary of Curious and Interesting Numbers*, is a number that a child wrote on the board at the kindergarten: 10000000000000000000000000000000000000000000000
00000000000000000000000000000000000000000000000
0000000. It is one followed by 100 zeros, the number the child considered the largest in the universe. The mathematician Edward Kastner, the uncle of the child who invented the googol, suggested that a much larger number might be called the googolplex, and that it be defined as 1 followed by a googol of zeros. A googolplex is thus 10^{googol}, which is a very large number indeed.[19]

This amusing game of naming larger and larger numbers can go on forever. We could propose the very large number defined by $10^{\text{googolplex}}$, or $10000^{\text{googolplex}}$, or a trillion raised to the power $1000000000000^{\text{googol}}$, or $\text{googolplex}^{\text{googolplex}}$, but even so we would never reach the "largest number." The reason is that there simply *is* no largest number. Given any number, all we would have to do is add to it the number one—giving us a greater number. The largest number does not exist; numbers go on and on infinitely far. To imagine this, just close your eyes and picture yourself flying in space. Ahead of you, you see numbers shooting towards you like the broken line dividing a highway: 1138, 1139, 1140 . . . 2567, 2568, 2569, 2570 . . . there are always more and more numbers, forever.

The genius of Georg Cantor was that he was among the first mathematicians (along with Bolzano, and possibly Galileo) to let his imagination run free and not be deterred by the concept of endlessness. Religious scholars who study the Kabbalah and Christian theology may have shared that courage in trying to imagine the unending immensity of the Divine. Cantor's Absolute and his transfinite numbers bear a resemblance to the image of God described by Augustine in *De Civitate Dei*. Augustine writes: "Every number is known to Him whose understanding cannot be numbered. Although the infinite series of numbers cannot be numbered, this infinity is not outside His comprehension. It must follow that every infinity is, in a way we cannot express, made finite to God."[20]

Cantor could take the fact that numbers run on forever without the uneasy feeling most people get from trying to imagine something we can not grasp, such as endlessness. Cantor accepted that infinity exists even without a way to see or touch or feel every one of these endless numbers, as we seem to want to do. He could accept *actual* infinity rather than the safe idea of a potential infinity of the Greeks and Cantor's contemporaries. Furthermore, Cantor was comfortable with the concept that there could be different *kinds* of infinity: one larger than the other—something most of us find too incredible to contemplate. Having made the psychological breakthrough, Cantor needed a language to describe these infinities. He wanted to give his infinities names.

Cantor called his infinities *transfinite numbers*, with the exception of the Absolute, which he viewed as beyond description. Until now, we've only seen two kinds of infinities: that of the integers, rational numbers, and algebraic numbers on the one hand, and the greater infinity of the transcendental numbers (and hence the entire real line containing them) on the other. Cantor knew one other order of infinity greater than the first two. This was the infinity of all functions, continuous and discontinuous ones, defined on the real line.[21] Cantor now needed to find a symbol to represent these varying orders of infinity he had discovered, so he could distinguish them from one another—and from any new ones he might encounter in his search.

At first, Cantor made an assumption about his transfinite numbers. He proposed the existence of a number that was infinite and yet the *smallest number greater than all the finite numbers*. Cantor reasoned as follows: we can generate the successive natural numbers by adding the number one—3=2+1, 4=3+1, 5=4+1, and so on. While there is no *greatest* number—you can always add one to any number and get a larger one—there is still the possibility of the existence of a number *larger than all the finite numbers*. Cantor named his first transfinite number ω, the Greek letter omega. If the number one was the "alpha," the first number, then the least infinite number greater than all the finite numbers was the "omega." Cantor then supposed that the principle of generation of numbers extended naturally to the transfinite numbers, allowing him to define the transfinite numbers $\omega+1$, $\omega+2, \ldots 2\omega, \ldots \omega^2, \ldots \omega^\omega, \ldots$, and so on. He now had an infinity of infinite numbers.[22]

Cantor soon went further, realizing that in order to develop his theory of the infinite he would need new definitions, and a new notation. Cantor was now in a position to extend the concept of a number using the ideas of set theory.

The cardinality of a set is a measure of the number of elements the set contains. For a finite set, the cardinality is simply the number of elements in the set. A set containing three dogs has cardinal number 3. A set of 106 people in a movie theater has cardinal number 106. What about the set of all integers? The set of all rational numbers? What is the cardinality of an infinite set? Cantor wanted to be able to define the cardinal number of an infinite set. At first, he used his ω to denote the cardinal number of a countable set such as the set of all integers. He also used the usual sign we often use to denote infinity: ∞. But soon he decided that the cardinal numbers needed new symbols. Cantor decided to name his infinities—his transfinite cardinal numbers—using the Hebrew letter aleph, $\aleph$. Why did he choose the aleph?

In the 1880s, when Cantor's first results were published, there was strong resistance to his ideas from mathematicians. Kronecker was the champion of the anti-Cantor camp, but other mathematicians were uncomfortable with the concept of an actual infinity as well. When this idea was further developed to include varying orders of actual infinity in Cantor's published work, these traditionalists became even more edgy, and a number of mathematicians began to fulminate against his heresy. But support for Cantor's work came from an unexpected source: the pope.

Pope Leo XIII made a strong attempt during his papacy to understand science and to bridge the apparent gap between scientific discovery and Church dogma. From the time he was elected Pope in 1878, Leo XIII inspired a transformation that resulted in a more enlightened atmosphere in the Church. The Holy See now encouraged and supported investigations of the laws of nature. Mathematics, and the role of infinity in particular, as seen in the light of Augustine's writings, gained importance during this period as a way to understand God.

The German priest Constantin Gutberlet held unusual views about infinity. Like Cantor, he believed that actual infinity could be contemplated by the human mind and that such contemplation could help believers get close to the Divine.

Gutberlet came under attack by other theologians and, in his defense, quoted Cantor's papers on infinity in his letters to rival theologians. Thus began Cantor's lengthy correspondence with the Christian clergy about the meaning of infinity and God. But Cantor had believed in the strong connection between infinity and God even before he began the epistolary exchanges with theologians in the late 1870s. Cantor had been brought up in a religious household. He always believed in God, and in later years—when under attack by mathematicians—used to say that God Himself had revealed to him the existence of the transfinite numbers. He knew the numbers were real because "God had told me so," and thus he required no further proof. But which God told him about the transfinite numbers?

Ivor Grattan-Guinness, a British historian of mathematics, argued in his paper, "Towards a Biography of Georg Cantor," that Cantor was definitely *not* Jewish.[23] Other biographers such as Boyer and Bell wrote that Cantor came from a Jewish background, and the letter from Cantor's brother reprinted in Charraud's book describes the family as being of Jewish descent on both sides. Grattan-Guinness, who seems to have had a desire to cast Cantor without any Jewish blood, still mentioned that the family was descended from Spain or Portugal several generations back, and that Cantor's wife, Vally, came from a Berlin Jewish family.

We know that the converted Jews of Spain and Portugal practiced their religion in secrecy for generations. The Marranos, forcibly-converted Jews of Spain and Portugal, became outwardly Christian, but at home practiced Judaism furtively, or at least followed some Jewish customs. Most of them knew Hebrew, and all knew their heritage. In private letters, such as the one written by Cantor's brother, they could be open with one another and freely discuss their families' traditions and provenance.

Grattan-Guinness suggested that the name Cantor was not Jewish and was derived from the Latin word for singer. It is true that the root of the word *cantor* comes from the verb to sing, but the particular form *cantor* rather than *chanteur* or *cantante* could only mean one kind of singer, the singer in the synagogue—the cantor.

The Spanish and Portuguese conversos left the Iberian peninsula in the fifteenth and sixteenth centuries and traveled north, first to Amsterdam and then farther east to Germany and Baltic Russia. It was probably in this way that Cantor's ancestors came to St. Petersburg, Denmark, and Germany. They kept their Jewish family name and their traditions, including that of marrying other Jews, as Georg Cantor had in fact done. While practicing Christianity on the outside, they were nonetheless well-versed in Judaism, and it was only through such a tradition that Cantor could learn the Hebrew alphabet. No other peoples in Europe studied Hebrew. They would have had no reason or inclination to do so.

The Iberian Jews also knew the Kabbalah. Jewish mysticism was born in Spain, and most of the important rabbis of the Kabbalah lived and worked in that country. As the Spanish and Portuguese Jews traveled and dispersed throughout northern Europe, they carried with them the tradition of Jewish mystical studies. The idea of the Ein Sof, the infinitude of God, was an important part of their communal hidden tradition. Georg Cantor had to have been aware of the role of the letter aleph as a symbol of God and His infinity. He told his colleagues and friends that he was proud of his choice of the letter aleph to symbolize the transfinite numbers, since aleph was the first letter of the Hebrew alphabet and he saw in the transfinite numbers a new beginning in mathematics: the beginning of the actual infinite.

But Cantor did not need to have had a thorough knowledge of Kabbalah to encounter the concept of infinity and its associated symbol, the aleph, within Jewish tradition. All he had to have been acquainted with was the most basic of Jewish prayers, *Adon Olam* (Master of the Universe). For a verse of this prayer, recited by Jews several times daily, is that God rules the universe *beli reshit, beli tachlit*: with no beginning, with no end. Hence the concept of infinity was known to anyone with a Jewish background, including the converted Jews of Iberia.

Cantor hypothesized the existence of a sequence of alephs. He named the lowest order of infinity, the infinity of the integers and the rational (as well as algebraic) numbers, aleph-zero. It is written as $\aleph_0$. Cantor believed that the alephs continued. He knew that the irrational numbers, in particular the transcendental ones, were more numerous than the rationals—they could not be put into a one-to-one correspondence with the rationals—and so there had to be a higher aleph to describe them. And the set of all functions on the real line had a still higher order of infinity, so a higher aleph would be required to describe them. But Cantor did not know whether there were any distinct orders of infinity, other alephs, *between* the aleph of the rationals and that of the irrationals, and between the one of the irrationals and that of the functions. This question would occupy much of the rest of his life.

At any rate, Cantor put forward the hypothesis that there was a series of alephs to describe higher and higher orders of infinity: $\aleph_0, \aleph_1, \aleph_2, \aleph_3, \aleph_4, \aleph_5, \aleph_6, \aleph_7. \ldots$

While he believed that larger and larger alephs existed, without knowing their exact placement, he did know something about the way the alephs interacted. Cantor discovered transfinite arithmetic. Here are some of the rules of this new mathematics of infinity that Cantor uncovered:

$$\aleph_0 + 1 = \aleph_0.$$

If we add one to the number representing the infinitely many integers, we still get that infinite number of integers (or rational or algebraic numbers). Adding one to the lowest order of infinity still leaves that infinity alone—it cannot take it to a higher level. Similarly, adding any finite number to the first order of infinity leaves that infinity the same:

$$\aleph_0 + n = \aleph_0.$$

Now, from Galileo's work showing that there are as many integers as there are squared integers (there is a one-to-one correspondence between the two sets), we know that adding two aleph-zeros to one another still leaves us with aleph-zero (adding the two infinite sets of the odd numbers to the even numbers gives us all the integers—a set whose cardinality is still aleph-zero). A similar argument could be that by adding all the integers (aleph-zero of them) to all the fractions (aleph-zero as well) we get all the rationals (still aleph-zero). So we have:

$$\aleph_0 + \aleph_0 = \aleph_0.$$

And similarly:

$$\aleph_0 \times n = \aleph_0,$$

where n is any number. Also:

$$\aleph_0 \times \aleph_0 = \aleph_0.$$

However, as we will see in the next chapter, the operation of exponentiation raises the cardinality of a set. Cantor was elated upon discovering the amazing laws of transfinite arithmetic. He now had a sequence of transfinite numbers, following their own logic and their own laws of arithmetic. If only he could show other mathematicians the truth and beauty of his divine theory, he would be a happy man. But, so far, it was only some theologians who appreciated his work. They understood that the mathematician had given them an elegant framework for understanding the immensity of God. Cantor had built a divine temple, level upon level of infinite numbers, the alephs. It would be hard to miss the uncanny similarity between Cantor's structure of alephs, each grounded on a previous one, and a visual image of the Kabbalah popular during that period. This is the picture of the nested circles of the Ein Sof—infinities also represented by alephs:

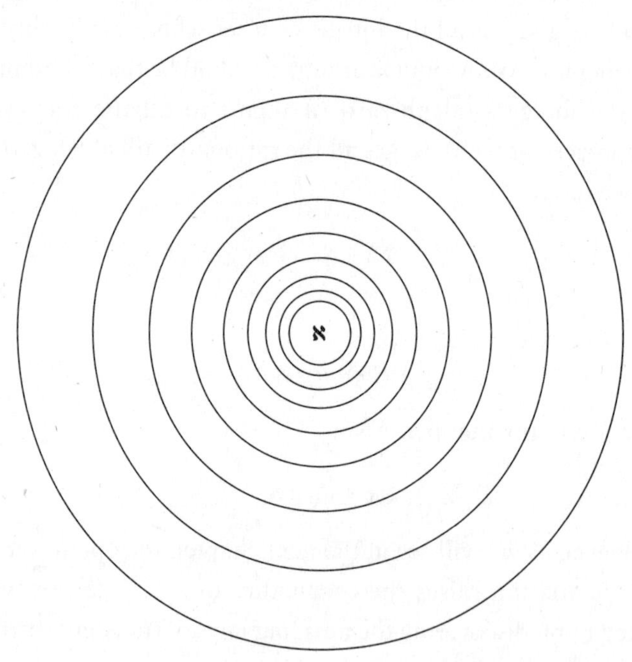

$\aleph_{12}$

The Continuum Hypothesis

Cantor was now eager to identify his alephs and the relationships among them. He had just opened the gate to the enchanted garden of the transfinite numbers—what he wanted now was to know their order. Cantor knew that the smallest order of infinity, the smallest transfinite number, was aleph-zero. He also knew, by an amazing theorem he proved in the 1870s, that for any set, whatever it may be, there is always a larger set: the set of subsets of the original set. For example, let's look at the set of three numbers {1, 2, 3}. What is the set of all subsets of this set of three elements? It is the set (called the *power set* of the original set) consisting of all possible subsets one could form from the set of three elements. These are: { }, the empty set; the three sets of one element each, {1}, {2}, {3}; then the sets of pairs, {1, 2}, {1, 3}, and {2, 3}; and finally, the set of all three elements of the original set, {1, 2, 3}. Thus, the power set of the set of three elements, the set of all subsets of the original set of three, has eight elements. The number of elements is obtained as $2^3=8$. In general, the logic is that every element in the original set has two possibilities: being included in a subset or not being included in the subset, hence there are $2^3=8$ possible subsets of a set of three elements.

Cantor knew that the set of all real numbers, the continuum of the real line, comprised all possible subsets of the set of all integers. Every integer can be included or not included in any of the infinite positions of a decimal number. Hence the number of elements of the continuum had to be 2 raised to the power of the infinite number of integers. The cardinal number of the continuum was therefore $c=2^{\aleph_0}$. There is an easier way of seeing this.

Every number on the continuum of real numbers has an infinite decimal expansion. Every number in the dense stretch of numbers on the line is represented by infinitely many (but countably so) integers from 0 to 9. In every one of the countably-infinite number of positions for a number there is one and only one digit: 0, 1, 2, etc. But we know that number systems using different bases are interchangeable. So every number on the real line can be written by a countably-infinite sequence of zeros and ones (that is, we write all our numbers to base 2). Therefore, for any given number, at every position there are two choices for the digit at that point: a zero or a one. There are aleph-zero such positions for every number. Hence, the number of numbers on the real line (the cardinal number of the continuum) is: $c=2^{\aleph_0}$.

The act of forming the power set—the set of all subsets—always produces a bigger set than the original, a set of a higher cardinality. Cantor could already see that his cardinal numbers were endless, since for any set he could produce there was a power set, and the power set had greater cardinality than that of the original set. So cardinal numbers got bigger and bigger as the sets representing them grew from set to power set to power set of the power set, and so on. As with the usual numbers, there seemed to be no end—no largest cardinal number. The problem of the existence of a largest cardinal number would appear again because of paradoxes inherent in set theory.

Similarly to the formation of the cardinal number of the continuum, *c*, by forming the set of all subsets of integers, one could obtain a set of still higher cardinality by considering the set of all subsets of numbers on the continuum, whose cardinality had to be: $d=2^c$. The process could continue onward, with no end. Cantor thus learned another property of his transfinite arithmetic: the operation of exponentiation does change the value of the alephs. Recall that:

$$\aleph_0 \times \aleph_0 = \aleph_0,$$

which means that even multiplying the cardinality of a countably infinite set by that of another countably infinite set still gives us countable infinity. For example, the set obtained as the product of the set of all integers by the set of all rational numbers still gives us a set whose cardinality is aleph-zero—a set that has the same "number of elements" as the set of all rational numbers. Also, recalling Cantor's "I see it but I don't believe it" discovery tells us that multiplying *c* by *c* still gives us the cardinal number *c* of the continuum. But exponentiation does change the cardinality of a set, and Cantor's transfinite arithmetic now had a new rule:

$$2^{\aleph_0}=c$$

or, in general,

$$n^{\aleph_0}=c,$$

where n is any finite number (2 is the smallest number for which this works).

But Cantor was not happy even though he had just discovered an entire new mathematics of the infinite. He wanted to know the *order* of his transfinite cardinal numbers. He wanted to be able to name them *consecutively* as:

$$\aleph_0, \aleph_1, \aleph_2, \aleph_3, \aleph_4, \aleph_5, \ldots \text{ and so on.}$$

In particular, Cantor asked himself the question: *Is there another cardinal number, another aleph, that lies between aleph-zero and the cardinal number of the continuum?* If the answer to this question was "no," then Cantor could then name the order of infinity of the continuum, c, simply $\aleph_1$. Not knowing the answer, he could not order his transfinite cardinals, because he couldn't tell which cardinal was indeed the next one after $\aleph_0$. Cantor could not name $\aleph_1$, nor any aleph higher than $\aleph_0$.

Cantor had a keen sense of what is important in mathematics. He knew right away that the question of the order of the alephs, and whether there were any other cardinal numbers—orders of infinity—lying between $\aleph_0$ (the lowest order of infinity) and c (the order of infinity of the continuum), was of paramount importance and would have far-reaching consequences.

Intuition might imply that the order of infinity of the continuum, the cardinal number c, should be the next aleph after $\aleph_0$. Since multiplication of alephs leaves them alone and exponentiation brings them higher, it made sense to Cantor that there should be no alephs between $\aleph_0$ and c. But he would have had to *prove* such a claim. And anyway intuition in mathematics often does not lead to the truth. Some mathematicians actually thought—and some still do today—that there are other alephs between $\aleph_0$ and c. But Cantor believed that c was the next aleph after $\aleph_0$. Thus he believed that $c = \aleph_1$. Since c was known to equal $2^{\aleph_0}$, what Cantor believed, and hoped to prove, was:

$$2^{\aleph_0} = \aleph_1.$$

This statement became known as the *continuum hypothesis*. This hypothesis about the orders of infinity is arguably one of the most important statements in all of mathematics. In 1908, this statement was generalized by Felix Hausdorff (1868–1942) to apply to all the alephs in the form of a *generalized continuum hypothesis*, linking every aleph with the one before it through the operation of exponentiation:

$$2^{\aleph_\alpha} = \aleph_{\alpha+1}.$$

Having written down the continuum hypothesis, Cantor set out to prove it.

Cantor had derived an entire theory of sets, and he naturally believed that his theory should give him a good basis for constructing the required proof. He knew the alephs had an order, and he couldn't see why it would not be the order he felt was logical for the transfinite numbers. Cantor spent all his time over a period of many years trying to prove the continuum hypothesis. The result seemed so natural to him, he was convinced a proof would soon follow.

On August 26, 1884, after working on the problem for several years, Cantor wrote to his friend and editor Gösta Mittag-Leffler. His letter described how elated he was that finally he had found an extraordinarily simple proof that "the continuum was equal in power to the second number class." (This was his way of describing in words the continuum hypothesis, $2^{\aleph_0} = \aleph_1$.) Cantor was rapturous about his discovery, and told his correspondent that he would soon send him the details of the proof.

But two months later, on October 20, Cantor sent Mittag-Leffler another letter saying that the proof he thought he had was a total failure. He was deeply depressed, feeling he was so close to his goal only to discover his proof was worthless. After recovering from the defeat, he set to work on a new approach.

On November 14, Cantor again wrote to Mittag-Leffler. In this letter, he told the Swedish mathematician a very surprising piece of news: he had just proved that the continuum hypothesis was *false*. Cantor now believed that the cardinal number of the continuum was not the second cardinal, $\aleph_1$. He was now convinced that he had a proof that the cardinal number of the continuum was above that of *any* of the alephs—there were infinitely many alephs between $\aleph_0$ and c.

This pattern of letters continued. Cantor kept changing his mind over and over again, thinking he had proved the continuum hypothesis, then thinking he had proved its converse. It made him uncomfortable to have to retract his findings and reverse the theorem in consecutive letters to his friend—a publisher on whose good will Cantor depended. Perhaps this embarrassment and his frequent bouts of depression made him withdraw a paper he had earlier sent to Mittag-Leffler. Consequently, Cantor never published in *Acta Mathematica* again.

Over the next few years, Cantor continued his touch-and-go relationship with the continuum hypothesis. After weeks of intensive work he would suddenly be convinced that he had found a proof of the theorem. Then he would find a fatal flaw in his derivations, and a few weeks later he would suddenly be sure that he had found a proof of the opposite result. Through this ordeal, and aggravated by Kronecker's continuing attacks, Cantor slowly went mad.

It might seem that if a mathematician couldn't be sure about his result—one moment thinking he had a proof of a theorem and the next thinking he had a proof of its opposite—then something was wrong with the mathematician, or at least with his logic. But today we understand that Cantor's predicament was not at all due to faulty reasoning. What Cantor didn't know—*couldn't possibly know*—was that he was working on an impossible problem. We know that the continuum hypothesis *has no solution* within our system of mathematics. Cantor felt alternately sure that the continuum hypothesis was true and that its converse was true because there really is no correct answer here. Both the continuum hypothesis and its converse are true. And both the continuum hypothesis and its converse are not true. The continuum hypothesis is undecidable within the realm of our mathematics. Unfortunately, the fact that we are unable to decide, unable to know, whether or not $2^{\aleph_0} = \aleph_1$ was discovered long after Cantor's death in 1918.

At the height of his efforts to obtain a solution to his continuum problem, Cantor suffered his first serious nervous breakdown. Was Cantor's fate very different from those of the rabbis of the second century who tried to enter God's secret garden, one losing his life and another losing his sanity?

ℵ13

Shakespeare and Mental Illness

It has been suggested that Cantor's illness might have been a bipolar disorder: manic depression. Some psychologists have even seen in his behavior hints of a persecution complex. E. T. Bell, writing in 1937, used a Freudian approach, tracing Georg Cantor's relationship with his father through the boy's childhood. Bell argued that the relationship between a sensitive boy who wants to please his parents and an authoritarian father who dictates to his child every aspect of behavior is bound to cause the child mental problems. Bell concluded that Cantor's depression was a result of the relationship with his demanding father.

Later scholars held that Cantor's bouts of depression and other symptoms were the result of his frustrating failures to prove the continuum hypothesis, exacerbated by the relentless torment unleashed on him by Kronecker. These experts discount the role of Cantor's father, although Nathalie Charraud—a psychologist and mathematician who studied the problem—does give some credibility to Bell's arguments.

Today we know that bipolar disorder is not simply "caused" by an event in the patient's life. There are genetic factors that contribute to the illness, and the environmental ones may not be as simple as frustration engendered by trying to solve a problem or the hurt brought about by criticism by one's peers. It is hard to judge the reasons for a mental illness even when the patient is at hand. To attempt to do so for a person who has been dead for the better part of a century, and without access to complete records of hospitalization and treatment, is much more difficult.

But modern psychology has found that depression can be caused when a person faces insurmountable difficulties. Whatever propensity toward mental illness Georg Cantor may have had, his failure to prove the continuum hypothesis and his anger and frustration over Kronecker's virulence had to have been a major cause of his condition. We know this both from the timing of Cantor's first mental breakdown and from the way he described his emotional state at the time to friends and colleagues.

Cantor's first collapse occurred immediately after he had withdrawn his paper from publication in *Acta Mathematica*, at the height of his frustrating attempts to solve the continuum problem. The depression was sudden and lasted two months, from May through June of 1884. On June 21, Cantor wrote again to Mittag-Leffler. He described his nervous breakdown and said he had recovered. However, he also expressed doubts about whether he would ever be able to return to mathematical research. His illness stultified him and made him doubt his abilities as a mathematician.

Over the following years, Cantor fell ill more and more frequently, and for increasing periods of time. When the attacks were severe, Cantor required hospitalization. A nervous bout usually began very suddenly, most commonly in the fall. Cantor's attacks began with paroxysms. He would rant about whatever enraged him at the moment: other mathematicians, German professors, scholars in other disciplines. Then depression would set in. The availability of medicines to control mental illness in those days was limited, and therefore the treatment was mostly symptomatic. Cantor was made to soak in a hot bath, participate in physical activity, and rest. When the symptoms abated, the doctors sent him home.

Cantor's first mental breakdown, in late spring 1884, differed from later ones: it was shorter, and it started in spring rather than fall. Cantor's eldest daughter, Else, was nine years old at the time of his first attack, and she was so profoundly affected by her father's behavior that she remembered it vividly and was able to describe it in detail to Cantor's first biographer, A. Schoenflies, years later.

Else and the other members of the family were shocked by the sudden change in Cantor's behavior. Cantor was clearly very deeply disturbed. He was withdrawn and could not work or interact with others. Following the attack, he had to spend many months resting and trying to regain his strength and emotional stability.

Cantor may have understood the forbidden nature of the knowledge he was seeking about infinity, for while he was recuperating he changed his mind about mathematical research and mathematics in general. While recovering from his nervous breakdown, Georg Cantor underwent a transformation. He became a Shakespearean scholar.

Cantor began to study English literary and historical problems, although his English was not perfect. He was fluent in German, and had a good knowledge of Danish and Russian, so English was his fourth language. Yet, once he had substantially recovered from the first attack of 1884, he spent all his time studying these subjects with one goal in mind. Cantor was determined to prove that Francis Bacon was the true author of Shakespeare's plays. It is not clear how or why a brilliant mathematician with limited knowledge of English and English literature thought he could achieve this goal. But Cantor kept pursuing a proof of this hypothesis.

Perhaps Cantor was looking for an escape from his serious problems with mathematics. If this was his reason, it might provide evidence for the hypothesis that his mental problems were produced by the impossibility of solving the continuum conundrum. But Cantor did not succeed in staying away from the continuum hypothesis. He did return to his efforts to solve the problem over the following years. Whenever he did so, he invariably got sick again and ended up at the Nervenklinik for several months. Rebounding from the ordeal, he would shift his attention to his efforts to prove the Bacon-Shakespeare hypothesis.

Recoiling from the continuum nightmare, Cantor tried to leave the department of mathematics. He asked the administration at Halle to transfer him to the department of philosophy at the University, but his request was turned down. At age 39, with his first attack, he gave up hope of ever obtaining a position in mathematics at the University of Berlin. Trying to switch to philosophy may have been his way of fleeing from his woes while still confined to Halle.

As the years passed, however, Cantor removed himself from teaching altogether. His mental attacks became more frequent and of longer duration. The University of Halle and the overseeing authority in Berlin were generous to Cantor, and patient with his problems. They granted him long sick leaves during his hospitalizations.

Now Cantor steadily collected works of Shakespeare and any information he could obtain about Shakespeare and Francis Bacon. In letters to associates, he described his great admiration for Bacon, and it would seem that it was this sentiment that made him eager to prove the unpopular theory that Bacon wrote Shakespeare's plays. Cantor tried to attend conferences, and in fact was given time to speak at scholarly meetings and express his views. Nathalie Charraud describes Cantor's work and his attack of mental illness following his frustrating attempts to prove the continuum hypothesis. Charraud believes that Cantor underwent a complete personality transformation after his first attack and, in a sense, felt that he himself had turned into a character in a Shakespearean tragedy.[24]

Charraud tells the curious tale of how Cantor came upon his strange new interest. While visiting an antiquarian in the nearby city of Leipzig, Cantor came upon an old book about Francis Bacon. Cantor assumed that the book was unknown to scholars. The book eulogized Bacon as a great poet rather than a man of science. Cantor took the descriptions as proof that Bacon was a great poet, and drew the conclusion that he must have written the Shakespearean plays. Tormented by years of harassment by Kronecker, Cantor empathized with the image of Francis Bacon he constructed in his own mind. Cantor identified with this distorted image of Bacon and embarked on the crusade to return to Bacon the acclaim he believed the scientist deserved for writing Shakespeare's plays.

In 1896 and 1897, Cantor published at his own expense two pamphlets arguing the theme. In one of these, he wrote: "Shakespeare is not immortal; on the contrary, it is Bacon who is immortal." Charraud goes further and draws an analogy between Cantor's desire to "expose to the world the true Shakespeare," and the wish he expressed to his colleagues to "expose to the world the true Kronecker."

As his obsession with Shakespeare and Bacon became more acute, Cantor's writings became stranger and more irrational. In 1899, after another failed attempt to solve the continuum problem, Cantor promptly suffered a new attack—as if God himself were once more punishing him for the crime of trying to understand the true order of infinities. Cantor was hospitalized in the Halle Nervenklinik for a lengthy period. He applied for medical leave for the academic year, and the ministry of education approved his request. When he was released from the hospital in the fall, Cantor asked for a leave of absence from the university, and his request was approved. Then he sent a curious missive to the ministry. In this letter he said he was eager to abandon altogether his professorship at Halle. He wrote that as long as his salary would not be reduced, he would be content to work in a library somewhere, where he could serve the Kaiser. He renounced his title of professor, and said that he would do anything to be released from the confines of a German university.

In the same letter, he boasted that he had great knowledge of history and literature and had published about the Bacon-Shakespeare issue. With the letter Cantor enclosed three of his pamphlets about the Bacon-Shakespeare problem, and nine of his visiting cards, on which he wrote his family history and made a reference to his "old beloved Czar Nicholas II of Russia." Cantor also wrote that he had great new proof concerning the true identity of the first king of England. "This will not fail to terrify the English government," he added. He requested that the Ministry send him an answer within two days, saying that if his request was turned down, he would apply to the Russian diplomatic corps, being Russian-born, and request to be of service to Czar Nicholas II.

The ministry of education seems to have ignored this letter Cantor wrote shortly after his release from a mental clinic. Cantor was not offered a position as librarian, nor did he join the Russian diplomatic corps in Berlin. He remained a professor at the University of Halle, but before too long he was again relieved of his teaching duties due to another hospitalization. The records at the university and the ministry of education show that the authorities made every effort to treat Cantor tactfully and, whenever possible, to grant his requests. He routinely obtained medical leaves and suffered no apparent antagonism from his supervisors even when, over their heads, he wrote bizarre letters to the authorities in Berlin. After his next release from hospital, Cantor was delivering a lecture on the Bacon-Shakespeare question in Leipzig when he received news that his youngest son, Rudolf, had died earlier that afternoon. The boy was a few days short of his thirteenth birthday.

Rudolf was frail and of poor health throughout his young life. He was a gifted violinist, and his family had great hopes that he would one day become a great musician. Following Rudolf's death, Cantor wrote letters to his friends filled with sorrow for the death of his son. He confided that he had himself regretted leaving music behind to become a mathematician, and said he had great doubts about his own choice of a career. Cantor was clearly disappointed in his choice—the frustrations of trying futilely to solve the continuum problem, and the ensuing mental attacks, made him unhappy and sorry for himself. He felt he had chosen a career that left him empty and unfulfilled. Now the hopes he and his family had held for their son were also dashed.

Surprisingly, following this terrible event as well as the death of Cantor's mother the same year, Georg Cantor managed to remain out of the mental clinic for several years. His next attack took place in 1902, and again the ministry of education promptly granted him sick leave for the duration. Cantor was in the hospital for the better part of a year. He was ill off and on for months, but emerged in time for the Third International Congress of Mathematicians, held in Heidelberg in 1904. At the congress, Jules C. König, a Hungarian mathematician, read a paper about the continuum. König's article argued that the cardinal number of the continuum was not any of Cantor's alephs.

Cantor was attending the Congress while recuperating from his latest bout of mental illness accompanied by two of his daughters, Else and Anna-Marie. On hearing König's paper, Cantor became outraged. He felt deeply humiliated to have his theory publicly berated, as he saw it. His feelings of persecution surfaced at once, and Cantor was unable to see in König's presentation a normal exchange of scholarly ideas that a congress such as this one was designed to promote. Once again, Cantor viewed himself as the lone protector of a truth under attack by sinister forces intent on quashing it.

According to Schoenflies, by 1904 Cantor had made the continuum hypothesis a matter of dogma.[25] He no longer required the proof that had eluded him for so many years. To him the assumption $2^{\aleph_0} = \aleph_1$ was not a statement that had to be proved. It was the word of God. Cantor described the continuum hypothesis to colleagues in exactly these words, and he even believed that God would protect the continuum hypothesis from its attackers.

When König's proof was read, the large audience of mathematicians present at the meeting seemed to accept that Cantorian assumptions—which by then had attracted much attention and promoted research by mathematicians—may have just collapsed. Cantor was devastated. He felt that God had betrayed him—He should never have allowed any errors to come to light in such a public, humiliating way. To pour salt on his wounds, the local press described König's finding as a sensational new discovery in mathematics. According to Joseph Dauben, another recent Cantor biographer, the apparent refutation of the continuum hypothesis, and other negative implications of König's paper on Cantor's work, was seen as so important that the Archduke of Baden hired a mathematician to explain to him the new results.

Cantor never believed a word of König's proof. To him

the continuum hypothesis was true. He was sure that there must have been a flaw in König's proof, and as distraught as he was after the attack on his work, he set out to find the error. König was a well-respected mathematician of great ability, and to find an error in his work would have been difficult. But Cantor immediately suspected that one of the lemmas—a preliminary result König had used to build his argument—was faulty. And indeed, less than a day after König's paper was read at the congress, the German mathematician Ernst Zermelo (1871–1953) proved that König had made improper use of the suspect lemma. That same year, 1904, Zermelo established his modern foundations of set theory, built on Cantor's ideas. A key hypothesis in Zermelo's theory was an enigmatic statement, the axiom of choice.

While the continuum hypothesis was not established by the refutation of König's theorem, and in fact there was no proof by Cantor or anyone else that $2^{\aleph_0} = \aleph_1$, Zermelo's argument against König saved Cantor's key assumption that the cardinal number of the continuum was indeed one of his alephs. It would have seemed that Cantor was saved—his theory was not demolished by König's paper after all. But Cantor was still upset. Ill and tired as he was, he was now becoming obsessed with König's paper. Cantor had to prove the continuum hypothesis before König would somehow repair his proof and proceed to destroy him completely.

Ironically, just as Cantor was becoming more delusional and losing whatever grasp he still had on reality, a group of mathematicians was forming around him, captivated by the power and elegance of his work and bent on saving Cantorian set theory. Zermelo was one of the key members of this group, and so was the famous German mathematician David Hilbert (1862–1943). Perhaps Cantor was finally triumphing over Kronecker's vicious attacks on his work, and getting the attention he deserved. But Cantor was not stable emotionally. His supportive circle of mathematicians had joined him at a resort hotel in Wengen following the congress. One day, Cantor woke up early and awaited his friends at the breakfast table. When they arrived, he greeted them—along with everyone else at the large dining room—with a loud, hysterical tirade against König.

The interest in Cantor's work had intensified over the previous two international congresses of mathematicians. Cantor attended the first congress, in Zurich in 1897, accompanied by his daughters Else and Gertrude, where they heard the virtues of set theory extolled by a number of leading mathematicians. Unfortunately, Cantor did not attend the Second International Congress of Mathematicians in Paris in 1900. For at that meeting, David Hilbert presented his now-famous "Ten Problems." (Later the set was expanded to twenty-three problems.) This was his list of unsolved mathematical conjectures which, he hoped, would be solved during the twentieth century. The first of Hilbert's "Ten Problems" (and the expanded group) was the continuum hypothesis.

Sadly, at a time when such prominence was finally being given to Cantor's unorthodox work on infinity, he was suffering periods of mental illness. Whenever he was ill, he spent inordinately long periods of time analyzing the Shakespeare-Bacon problem. By the early 1900s, he had amassed an entire library of books and papers on the works of Shakespeare and on Bacon's life and ideas. In 1911, while enjoying a period of relatively good health, Cantor realized a life dream—he visited Britain, the home of Bacon and Shakespeare. Back in 1908, Cantor had promised a British mathematician to send a mathematical paper to the *Journal of the London Mathematical Society*, but never wrote the paper. The invitation to come to Britain, however, remained, and in September 1911 he came to St. Andrews University in Scotland as an invited Distinguished Foreign Scholar. The invitation was to the department of mathematics, with the expectation that Cantor would discuss set theory and his work on infinity. When he arrived, Cantor behaved erratically. And instead of mathematics, he discussed the Bacon-Shakespeare issue, to the surprise and embarrassment of his hosts. Then Cantor unexpectedly left for London.

From London, Cantor wrote to the great English mathematician and philosopher Bertrand Russell (1872–1970). Russell had just finished his seminal volume coauthored with A. N. Whitehead, the *Principia Mathematica*. Since the work attempted to lay a foundation for the entire field of mathematics based on set theory, Cantor was eager to meet Russell. Cantor sent Russell confused letters, containing words continuing into the margins and lines written from top to bottom across lines written from left to right. Cantor wrote two such letters, but the two men never met.[26]

In his *Autobiography*, Lord Russell chose to publish the letters. He described Cantor as one of the greatest intellects of the nineteenth century. But then he added insensitively: "After reading the following letter, no one will be surprised to learn that he (Cantor) spent a large part of his life in a lunatic asylum."[27]

We can only guess what might have happened had Russell actually met Cantor in 1911, for Russell's work on the foundations of mathematics, including his famous paradox, were important for the later development of Cantorian set theory and concepts of infinity.

$\aleph_{14}$

The Axiom of Choice

Cantor realized that if he ever hoped to prove the continuum hypothesis, he had to establish a way of comparing his transfinite cardinal numbers. Doing so would establish that every transfinite cardinal was a member of the system of alephs, and thus there was no cardinal number outside the ordering: $\aleph_0$, $\aleph_1$, $\aleph_2$, $\aleph_3$, $\aleph_4$, $\aleph_5$,. . . . Cantor gave his sequence of alephs the name taf, ת, the last letter in the Hebrew alphabet. He did so to imply finality: every infinite cardinal had to be an aleph—had to belong to the system ת containing all alephs. There were no infinite cardinals outside his system, although the system went on forever—there were always greater and greater alephs.

But before Cantor could prove that every infinite cardinal number had its place within the system ת, Cantor needed a way to compare every possible pair of cardinal numbers. The infinite cardinals had to have the same ordering principle that the real numbers on the line had: namely, that for any two of them, either they were equal ($a=b$) or one was greater than the other ($a<$b or $a>b$). To achieve this property for the transfinite cardinal numbers, Cantor had to define a particular property of sets. We call this property *the well-ordering principle.*

The well-ordering principle says that every set can be well-ordered. And a set is called well-ordered if every one of its non-empty subsets has a smallest element. Let's look at an example. If our set is {1, 2, 3}, then we know that the set of all subsets has eight elements ($2^3=8$, as we've seen earlier). One of these subsets is the empty set, and the other seven are: {1}, {2}, {3}, {1, 2}, {1, 3}, {2, 3}, {1, 2, 3}. The original set, {1, 2, 3} is well-ordered because each of its non-empty subsets has a smallest element. These smallest elements are (in order): 1, 2, 3, 1, 1, 2, 1. Cantor needed to prove the well-ordering principle, namely to prove that *every set* (in particular, infinite sets) could be well-ordered as in the example above.

If he could achieve this task, he could then show that every transfinite cardinal number had to be one of his alephs. If he couldn't achieve this task, there would be no hope of ever proving the continuum hypothesis, since it would then be possible that the cardinal number of the continuum, *c*, was something other than an aleph. If *c* was something other than one of the alephs, $\aleph_0$, $\aleph_1$, $\aleph_2$, $\aleph_3$, etc., then there would be no way of placing it within this *ordered* set of cardinals, a prerequisite for showing—with much luck—that it was indeed the *second* transfinite cardinal number in the system ת, that is, $\aleph_1$.

Cantor was unable to prove the well-ordering principle, his prerequisite for any important headway on the continuum hypothesis. Then, in 1904, König dealt him the frightening blow of supposedly proving that *c* was not one of the alephs. Although Zermelo saved Cantor the next day by exposing the flaw in König's proof, Cantor's work was now in danger of further assault. The shaken Cantor realized more than ever that he now desperately needed a proof of the well-ordering principle. Zermelo, who had come to Cantor's rescue, continued his work in an effort to help him. And where Cantor had failed, Zermelo was successful. Still that same year, 1904, Zermelo produced a proof of Cantor's well-ordering principle.

Ernst Zermelo was born in Germany and attended the University of Berlin, obtaining a doctorate in mathematics in 1894 with a dissertation on the calculus of variations. He became a professor at the University of Zürich but had to resign the position after a few years because of poor health. In 1926, Zermelo was appointed an honorary professor of mathematics at the University of Freiburg in Germany. When the Nazis gained control in Germany, Zermelo was one of the few academics who resigned their positions in protest against the regime.

In 1904 Zermelo was just beginning his life's work—formalizing Cantor's set theory. The work commenced with his attempts to prove Cantor's well-ordering principle. Zermelo also realized that the continuum hypothesis could only hold if every infinite cardinal number was one of Cantor's alephs, and that to prove this necessary condition, one had to show that every set could be well-ordered. After working on the problem for much of the year, Zermelo finished the proof on September 24, 1904, and was even successful in providing an actual method of obtaining an ordering of any set.

The proof began by associating a *representative point* with every non-empty subset of a given set. Such a point he called a "distinguished element" of the subset. The representative point from each subset is simply chosen from among all the points of the subset. To carry out this choice of a single element from every subset, Zermelo relied on a selection principle, which he called the "axiom of choice." His proof of the well-ordering principle had simplicity and elegance, of which he was proud.

But within days of the publication of Zermelo's proof, a number of mathematicians raised serious objections to it. The problem with the proof was exactly the added axiom Zermelo had used—the axiom of choice. In a finite world, making a choice is a simple procedure. Not so once we enter the realm of infinity. Even in the simplest hypothetical case, where every subset has only two elements, once there are infinitely many such subsets, what is there to guarantee us the ability to choose? The problem mathematicians saw here was that Zermelo could not prescribe a *way* for making a choice infinitely many times. Just saying that such a choice could be made was not enough. These mathematicians wanted an exact rule stating *how* such an infinite sequence of selections could be made. Immediately, the axiom of choice (and Zermelo's proof of the well-ordering principle, which depended on it) became suspect.

The controversy about the axiom of choice never did subside. Over the years, mathematicians have found that a number of mathematical principles were equivalent to the axiom of choice. Many mathematicians stay away from all of these equivalents as well as from the axiom of choice itself. Proofs that require the use of the axiom of choice are considered questionable, and mathematicians often seek alternative proofs that do not rely on the principle of making a choice

THE AXIOM OF CHOICE

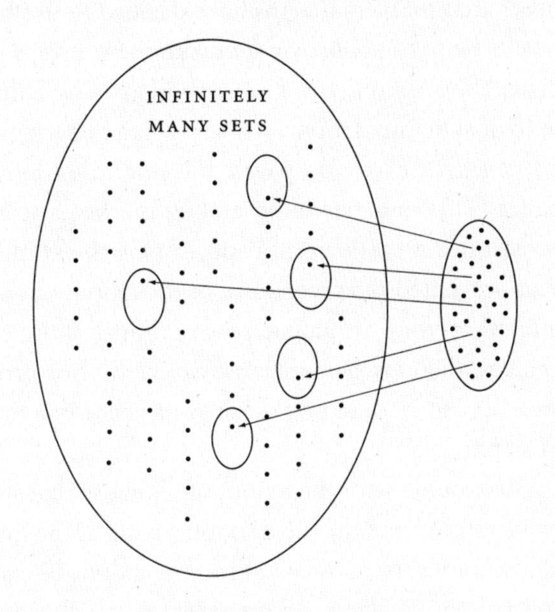

A SINGLE POINT IS SELECTED FROM EACH ONE OF THESE SUBSETS OF THE LARGER SET

infinitely many times. The theorem that Zermelo had set out to prove—the well-ordering principle—was shown to be one of the equivalents of the axiom of choice. Thus his result was itself suspect, since believing the proof was now tantamount to believing that one could, indeed, make a choice of an element from a set infinitely many times.

Mathematics is founded on a set of axioms. These axioms are thought to be self-evident statements, and subsequent developments hinge on them. From the axioms, together with rules of logic, propositions or theorems are established by rigorous proof, each supporting the other. The method of proof that mathematicians have decided to use requires that results be obtainable by a *finite* number of steps. If I want to prove that A implies B, I must show it using a finite number of logical manipulations. My proof can be written on a page, 20 pages, or even 300 pages, but not an infinite number of pages. This understanding among mathematicians of what constitutes a mathematical proof was the stumbling block that prevented the acceptance of the axiom of choice. If you infer that your argument relies on an infinite sequence of choices, then mathematicians may view your proof as problematic: How could you finish your proof in a finite number of steps?

The trouble with the axiom of choice is that it is different in its nature from the other axioms lying at the foundation of mathematics. In its very nature, the axiom of choice is nonconstructive. There is no prescription to tell us how the infinite sequence of choices is to be made. The question of whether or not the axiom should be accepted has become the most hotly contested issue in the foundations of mathematics.

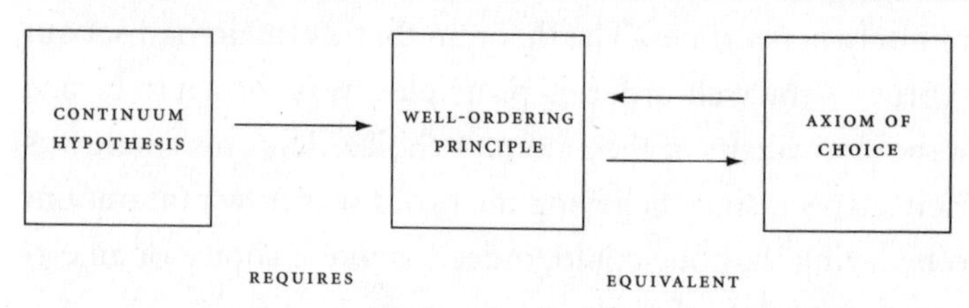

A proof of the continuum hypothesis would require the well-ordering principle. Any consideration of the continuum of numbers on the real line cannot escape the fact that these numbers are *ordered* from small to large. Consequently, there is no theoretical way to look at these numbers without considering this property of order, i.e., it makes no sense to try to jumble up the numbers and consider them as an unordered set. Because of this inescapable dependence of the numbers in the continuum on their order, any meaningful analysis of the continuum must proceed with the aid of the well-ordering principle. But the crucial principle of well-ordering is itself equivalent to the axiom of choice. Therefore, from the moment the well-ordering principle was invoked, Cantor's elusive continuum hypothesis and Zermelo's axiom of choice were forever intertwined. Within two decades, Kurt Gödel would shock the world of mathematics by proving a striking property of the axiom of choice and the continuum hypothesis. But even before that, the paradoxes would hit hard.

By the time Zermelo began his pioneering work in an effort to establish Cantor's continuum hypothesis, formalizing Cantor's entire theory of sets, Cantor himself was sliding away from the world of rigorous mathematics. The idea of a mathematical proof was becoming less important in his mind. His role, as he saw it, was that of a faithful secretary to God.[28] He believed he was entrusted with the task of recording God's words to the world. The continuum hypothesis was a statement about infinity—the realm of the divine—and Cantor was the intermediary through whom these results were to be communicated to the world. As his mental illness progressed over the following years, Cantor spent more and more hours by himself, sitting in a room at his home or in the hospital, staring at a blank wall and meditating. These meditations were no different in their essence from those of the Kabbalah. Both Cantor and the Kabbalah practitioner meditated on the infinitude of God; both knew they were entrusted with a deep truth; and both felt that no proof was necessary.

$\aleph_{15}$

Russell's Paradox

There is an ancient paradox attributed to the sixth century B.C. philosopher Epimenides the Cretan. The Cretan says: "I am lying." Should you believe him? If his statement is true, then he is lying and the statement is false. If the statement is false, then he is not lying and the statement is true. In a letter to Titus, the Apostle Paul refers to this ancient paradox by saying: "All Cretans are liars, one of their own poets has said so."

Another paradox of antiquity is the dilemma of the crocodile. A crocodile steals a child and then says to the child's father: "I will return your child to you if you can guess correctly whether or not I will return the child." The father replies: "You will not return the child." What should the crocodile do?[29]

In 1897, the Italian mathematician Cesare Burali-Forti (1861-1931) discovered a paradox inherent in Cantor's theory of sets. Burali-Forti considered the entire succession of *ordinal* numbers. He noticed that this set had to contain an ordinal number greater than all ordinal numbers. But by definition the set of ordinal numbers must contain all ordinal numbers and to every such number one could be added. Therefore, there can be no set containing all ordinal numbers. By an extension of Burali-Forti's idea, it can be shown that there can be no greatest aleph. Cantor was aware of Burali-Forti's paradox in 1897, as he mentioned it in a letter to another mathematician. In Zermelo's foundation for set theory based on Cantor's work, the problem raised by Burali-Forti is solved by assuming that the set of all ordinal numbers simply does not exist. A better-known paradox, which also complicated the development of Cantor's ideas, was the famous Russell's paradox.

Bertrand A. W. Russell was one of the most famous philosophers of the twentieth century. His writings on political freedom helped him earn the Nobel Prize for literature in 1950. Russell was a well-known pacifist and contributed to metaphysics, epistemology, ethics, and other areas. Russell's contributions to mathematics include his famous work, the *Principia Mathematica*, which he co-authored with Alfred North Whitehead, consisting of three volumes published between 1910 and 1913. These tomes were designed to establish a perfect logical foundation for all of mathematics. The volumes became the basis for the work of a number of important logicians who were laboring to establish a logical, complete theory of mathematics. Among them was Kurt Gödel, who later surprised the world of mathematics by showing that the edifice constructed by Russell and Whitehead was far from adequate as a foundation for mathematics.

But Bertrand Russell is equally famous in mathematics for proposing a disturbing paradox that would forever plague mathematical logic: Russell's paradox. It became the best known of all the paradoxes of set theory. Russell's paradox can best be illustrated by a famous analogy popularly known as the story of the barber of Seville. The barber of Seville shaves all the men in the city of Seville who do not shave themselves. Now comes the obvious question: Does the barber of Seville shave himself? If he does, then he doesn't. If he doesn't, then he does. This is a logical paradox.

A semantic example of Russell's paradox, called Grelling's paradox, is the following. A predicate may be true of itself, or (more often) it may not. For example, the word "short" is indeed a short word. The word "English" is indeed an English word. The word "word" is indeed a word. The word "pentasyllabic" has, itself, five syllables. But the word "long" is a short word; "German" is not itself a German word; and the word "monosyllabic" has five syllables and not one. The paradox materializes when we ask whether "not-true-of-self" is true of itself or not.[30]

In set theory, the actual Russell's paradox deals with sets. Sets may contain other sets as members. A basket containing fruit can be viewed as a set of elements: the basket is the set, and its members are the fruits inside the basket. But let's now look at another set: the set of all baskets of fruit on the ground at a picnic. This is a set whose elements are sets of fruit. Also, some sets contain themselves as members. For example, the set of all things in the world that are not dogs contains itself as a member. This is true because this set is not a dog.

In 1901, Bertrand Russell asked himself a seemingly simple question, which shook set theory and the structure of logic underlying mathematics. Russell considered the set of all sets that are not members of themselves. Russell called this set *R*. Then he asked: Is *R* a member of itself? Here, Russell obtained a paradox. If the set *R* is a member of itself, then it isn't. And if *R* is not a member of itself, then it is.

Russell sent a description of his paradox to the German logician F. L. G. Frege (1848–1925), who had just developed an alternative system of axioms as the basis for the theory of sets, and was about to publish his work. Frege's axioms would play an important role in the development of set theory. Axiom V in the system Frege proposed in 1893 was the crucial axiom of abstraction. According to this axiom, given any property, there exists a set whose members are just those elements that possess the property. Russell's paradox shot a hole through Frege's axiom V, for here was a property that could not be satisfied by any set because it leads to a contradiction.

When he got Russell's letter, Frege had to go back to the drawing board and try to come up with another axiom. Two years later, in an appendix to the second volume of his work, he recalled his reaction upon receiving the news from Russell about the paradox: "Hardly anything more unfortunate can befall a scientific writer than to have one of the foundations of his edifice shaken after the work is finished. This was the position I was placed in by a letter of Mr. Bertrand Russell . . . Even now I do not see how arithmetic can be scientifically established; how numbers can be apprehended as logical objects, and brought under review; unless we are permitted—at least conditionally—to pass from a concept to its extension."[31]

A key implication of Russell's paradox was that there can

be no universal set, or set that contains everything. In the early approaches to set theory, the existence of a universal set containing everything was taken for granted. Russell's paradox showed that it is impossible in mathematics to obtain something for nothing. It is not enough to *define* a set simply by decree. We must also have at hand a set whose elements actually exist. By defining a unicorn we do not prove that unicorns actually exist. Set theorists, recognizing the shaky grounds on which they stood due to Russell's paradox, were left with the task of trying to determine which properties actually define real sets. But there is still no known way to achieve this aim. As Gödel was to prove within two decades of Russell's work, a complete answer to this question may not be possible.

Russell's paradox and its implications were important hurdles on the way to an axiomatization of set theory and the foundations of mathematics. The paradox, in different forms, had been known earlier. Zermelo himself is said to have discovered what we call Russell's paradox, but never thought of publishing it on his own. To him the paradox was obvious.

An interesting paradox in the foundations of mathematics is brought about by the axiom of choice. It is called the Banach-Tarski paradox, after its discoverers, the Polish mathematicians Stefan Banach (1892–1945), who introduced advanced vector spaces into mathematics, and Alfred Tarski (1901–1983), who did pioneering work in logic. The Banach-Tarski paradox starts with an application of the axiom of choice. By mathematical derivations in Euclidean space (the usual space of three or more dimensions in which geometry is studied), the two mathematicians have shown that a sphere of a fixed radius can be decomposed into a finite number of parts and then put together again to form *two* spheres each with the same radius as the original sphere. This paradox came as a big surprise to mathematicians. By accepting the axiom of choice, mathematics allowed for the existence of something that seems like pure magic.

The Banach-Tarski paradox:

$\aleph_{16}$

Marienbad

Georg Cantor was discouraged by the paradoxes which proliferated in the early years of the twentieth century. Russell's paradox and its many relatives discovered by other mathematicians, both before Russell and after him, threatened the very foundations of mathematics. Cantor's continuum hypothesis is intimately related to concepts that lie deep within the foundations of mathematics, and therefore the paradoxes made it less likely that Cantor would ever overcome his difficulties and prove the continuum hypothesis.

As the years went by, Cantor fell deeper into despair, and his episodes of mental illness became more frequent. He required longer periods of hospitalization in the Halle Nervenklinik. During World War I, most patients were removed to an alternate facility to make place for soldiers who had to be accommodated in the city. Only two patients remained in the Nervenklinik during this time: the wife of a wealthy judge, who stayed there for eleven years because her family did not want her moved to an insane asylum, and Georg

Cantor. Cantor's general health was clearly in decline. What remained constant in his mind was an unwavering belief that God gave the world the continuum hypothesis through him. As he was becoming detached from the real world, Cantor's mind drifted into a state where reality and fantasy could no longer be distinguished.

During this same period, a bright child was growing up in Czechoslovakia. His name was Kurt Gödel, and his family circumstances could not have been more different from those of Cantor. As fate would have it, Gödel would become Cantor's successor, and would prove to be one of the greatest minds of all time.

Kurt Friedrich Gödel was born on April 28, 1906—in the middle of the turbulent years in set theory between the time of König's presentation in 1904 and Zermelo's axiom of choice of 1908—in the city of Brno in Moravia, Czechoslovakia (an area now part of the Czech Republic). The Gödel family was ethnically German, and its members had lived in Czechoslovakia and Austria for generations and had strong ties to Vienna. Gödel had one brother, Rudolf, four years older, to whom the young Kurt was very attached.

Kurt's father, Rudolf senior, did very well in business during the early years of Kurt's childhood, and within a few years was able to build a three-story house for his family. Two aunts also lived in the house with the Gödels for many years. Kurt's mother, Marianne (Handschuh), came from a higher social class than that of her husband and was well-educated. She raised the two boys in hopes they would become refined Austro-Hungarian gentlemen, versed in the arts, music and languages. Kurt grew up speaking a number of languages—but he rarely spoke Czech, the language of the land, since his family considered it less refined than German.

Ethnic Germans constituted about a quarter of the population of Czechoslovakia at that time, living mostly in cities, and mostly in proximity to the borders of Germany and Austria. In the cities the ethnic Germans inhabited, the German language was widely spoken. The Gödels and many other ethnic Germans employed Czech servants, and considered the Czechs less educated.

The Gödels' house had a garden with fruit trees, which was large enough for the exercise of their two dogs.[32] As a child, Kurt Gödel played quietly with his brother, read, and asked questions. Early on, he showed an insatiable curiosity about the world, constantly asking his parents "Why?" The family affectionately nicknamed him "Der Herr Warum"—Mr. Why. At the age of ten, Kurt enrolled in the gymnasium, the Austro-Hungarian educational institution where boys (and few girls) spent their years studying the classics as well as science and mathematics in preparation for a university education. Surprisingly, the boy destined to become one of the greatest mathematicians of the century received a grade of "very good" in every subject he took except for mathematics, where his grade was "good" on a report card of 1917.[33]

The Gödels seem to have been oblivious to developments in the world around them. In 1914, World War I erupted. The Gödels lived close to the center of the violence, but there is no evidence that the war disrupted their life in any way. The father continued to develop his thriving business; the mother continued to enjoy the cultural life in Brno and

Vienna; and the two boys continued their daily routines of schoolwork and recreational activities. The family took frequent holidays at fashionable resorts, a habit that was not disrupted by the war.

In June 1917, a mere 250 miles northwest of Brno, in Halle, Georg Cantor was admitted to the Nervenklinik for the last time. Food shortages and the privations of war made life in the mental clinic difficult and unpleasant. Cantor, now in his seventies, did not want to go into the hospital. He begged his wife and the doctors to allow him to stay home, but they ignored his requests. By the end of the year, he sent his wife the last forty pages of his daily calendar as proof that he had lived through the end of 1917. But on the sixth of January, 1918, he was found dead in his bed. The reported cause of death was a heart attack, but Cantor was by then very thin and evidently had eaten little over a period of months.

Cantor's final legacy, beyond the discovery of the transfinite numbers and the continuum hypothesis, was his realization that there could be no set containing everything (as Russell's paradox implied as well) since, given any set, there is a larger set—its set of subsets, the power set. There is thus no largest cardinal number—the Absolute is beyond our reach. Identifying this concept with God, Cantor wrote in one of his last letters to the English mathematician Grace Chisholm Young:

"I have never proceeded from any 'Genus supremum' of the actual infinite. Quite the contrary, I have rigorously proved that there is absolutely no 'Genus supremum' of the actual infinite. What surpasses all that is finite and transfinite

The Halle Nervenklinik. *Credit: Amir D. Aczel.*

is no 'Genus'; it is the single, completely individual unity in which everything is included, which includes the *Absolute*, incomprehensible to the human understanding. This is the *Actus Purissimus*, which by many is called *God*."[34] Perhaps Cantor's ultimate belief that the absoluteness of God was incomprehensible to the human mind—even when the mind can attempt to understand actual infinity—brought peace to his tormented soul.

In Brno, the budding young genius of Kurt Gödel continued its development during this difficult period, sheltered in the privileged milieu of the well-to-do. Kurt read widely in a variety of fields and took advantage of the resources of the gymnasium. When he was fifteen, not long after the Great War, the family once again went on holiday. This time they spent a few weeks at the famous spa of Marienbad, in nearby Bohemia.

Marienbad: the name conjures up images of expensively dressed men and women walking leisurely on wide white paths though expansive manicured gardens, large fountains spewing the mineral-rich waters high into the air. . . . The Gödel family is likely to have stayed at the elegant Baroque-style hotel at the springs, where many famous people have enjoyed their holidays, among them King Friedrich Wilhelm IV of Prussia, King Otto I of Greece, the Persian Shah Nasreddin, Edward VII of Britain, as well as Goethe, Mark Twain, and Sigmund Freud, to name but a few.

As Kurt described the experience many years later, at Marienbad he underwent a transformation. Until then, Gödel expected to pursue his interests in the humanities, social studies, and languages, as an educated man of the period. But walking the long corridors of the elegant hotel, strolling through the lavish parks, and soaking in the steaming mineral waters, he was suddenly changed. A mysterious force was drawing him into a strange world of equations, symbols, and infinity. By the time the family returned home from vacation, young Kurt was a mathematician.

ℵ17

The Viennese Café

Kurt graduated from the gymnasium in 1924, and moved to Vienna to enroll at the university. In his last three years at the gymnasium, he had concentrated on mathematics but shown great interest in philosophy and physics as well. These three interests stayed with him throughout his life. He devoted many years to uniting concepts from the physical world with philosophical ideas and with the mysterious discoveries he made in the foundations of mathematics.

Kurt had done extremely well at the gymnasium, and could have attended any university in Europe. Berlin, with its impressive concentration of leading mathematicians, certainly was an attractive place for the bright young student. But Kurt was very attached to his family. He therefore decided to go to the nearest prestigious school: the University of Vienna.

The distance from Brno to the Austrian capital is less than seventy miles, so Gödel was close to home when in Vienna. The city, too, held great attractions for him. It was the nearest large city to his hometown, German was spoken there, and there were many pleasures to be pursued in this center of culture, art, and music. In addition, Kurt's older brother was already there, attending the university. Kurt arrived in Vienna and moved in with his brother to share a small apartment not far from the university. Rudolf took care of his younger brother, who was already occasionally suffering from the mild ailments—part real, part imagined—that would plague him throughout his life.

Kurt started taking courses in mathematics, and was moved by the lectures of some of the well-known professors at Vienna. One of them was the mathematician Hans Hahn (1879-1934). Hahn received a doctorate in mathematics from the University of Vienna in 1905. He served in World War I and was seriously wounded in battle. When he recovered, Hahn went to the University of Bonn, where he was appointed a full professor of mathematics. It appears that Hahn missed Vienna, and in the summer of 1921, while Kurt and his family were soaking in the waters of Marienbad, Hahn returned to his home city and took up a professorship at the University of Vienna. He made contributions to many areas, but his most seminal result in mathematics was the famous Hahn-Banach theorem, attributed jointly to him and to Stefan Banach.

The Hahn-Banach theorem is in the area of functional analysis. The theorem gives conditions under which a linear functional (a mapping of a vector space into the real line, which is both linear and homogeneous) can be extended to the full space that shares boundedness conditions with the functional.[35] The Hahn-Banach theorem is very important in advanced mathematical analysis. Its proof has a mysterious quality—it requires the use of a result called Zorn's lemma. Zorn's lemma says that if every chain in a partially ordered set has an upper bound, then the partially ordered set has a maximal element. But Zorn's lemma is equivalent to the axiom of choice. The Hahn-Banach theorem is an example of an important result in mathematical analysis whose logic relies on an assumption at the very foundations of mathematics. The proof of the Hahn-Banach theorem relies on whether or not the axiom of choice holds.

As Gödel continued on to do doctoral work in mathematics, Hans Hahn became his dissertation adviser. Through studying mathematical analysis and the major theorem proved by his adviser, Gödel came to understand the power and importance of set theory and the foundations of mathematics. He proceeded to study the mysterious axiom of choice and the properties of infinity underlying it. He also studied Cantor's continuum hypothesis. It is interesting that both Gödel and Cantor derived their interest in set theory and the properties of infinity from the study of problems in mathematical analysis.

Hans Hahn had wide-ranging interests. One of them was the occult. He believed in seances and in the possibility that the spirits of the dead might speak to the living. It is not clear how vigorously Hahn pursued such non-mathematical interests, but he liked to discuss them with his friends in cafés.

Hahn organized a circle of his friends, among them his student Gödel, most of them mathematicians and philosophers or scientists from various fields. The meetings at the Viennese cafés became more regular in 1924, and their participants called themselves the Circle of Vienna, or, sometimes, the Schlick Circle, after the philosopher Moritz Schlick. The meetings were lively, and their topics ranged over many areas of science and philosophy. Members drank coffee, played backgammon, took walks together, and discussed ideas. Young Gödel took part in almost every café meeting, and his colleagues later described him as very quiet, intently listening to what others had to say, often nodding.

Here in the intellectually stimulating milieu of Vienna's cafés Gödel developed the ideas of his famous dissertation and subsequent papers, which forever changed the face of science. Once directed toward the foundations of mathematics and Cantor's work on actual infinity, Gödel used his deep philosophical bent to ask key questions: What is proof? Is proof equal to truth? Is something true always provable? And: can a limited system yield proof of something that extends beyond the system?

Gödel worked hard, but he also played hard, and during this period of intense investigations he met an attractive dancer six years his senior, Adele Porkert. Within a few years, despite his family's objections, the two were married in a private ceremony in Vienna. Between classes at the university, partying at night with Adele in Vienna's nightclubs, and spending afternoons and evenings in cafés with the Schlick Circle, Gödel found time to write his stunning dissertation.

Given any system, Gödel concluded, there will always be propositions that cannot be proven within the system. Even if a theorem is true, it may be mathematically impossible to prove. This is the essence of Gödel's famous incompleteness theorem.[36] The human mind, existing within a limited universe, cannot perceive an immense entity that extends beyond the confines of the system.

Gödel's theorem is somewhat related to Cantor's theorem about the nonexistence of a largest cardinal number. Cantor had shown that given any set—however large it may be—there exists a larger set: the set of all subsets of the given set. Given any infinite system, there is always a larger infinite system, one whose cardinality is greater. Within *any* limited system, there exist entities that cannot be perceived or reached or proved, and we need to move up to a larger system in order to comprehend these entities; but when we do this, we encounter larger systems and entities that lie beyond them. This concept can be illustrated using the Russian doll analogy:

Is there a *largest possible* Russian doll—one that contains all the other Russian dolls? This question is closely related to Russell's paradox and to Cantor's idea that, given any set, there is always a larger set. The impossibility of a set containing everything brought Cantor to the conclusion that there was an *Absolute*—something that could not be comprehended or analyzed within mathematics. Cantor identified the Absolute with God.[37] The Absolute can be identified with the Kabbalists' *Ein Sof*—an infinity so great that it lies outside what humans can comprehend. The Russian doll principle, the impossibility of a universal set, and the unattainable Absolute perhaps lend credibility to Gödel's incompleteness principles: there is always something outside, something larger than any given system.

We can see how a system may be incomplete with respect to some theorems, unprovable within the limited system, by analogy with the operating system of a computer. Suppose that you are working on a document on your computer screen. You can do many things: write, move text, cut and paste information, even insert art or formulas from other applications. But you can't delete the document on which you are working from within that document itself. To do this (or to move the document to another file), you need to exit your document and carry out the operation from within a larger system.

Some ideas or properties are not seen or perceived within any given system, and in order to understand them we need to transcend to a higher level. Since there is no "highest" level—as Cantor has shown—there will always be ideas or properties that we can't understand or address within any system. By analogy, a mere human can never attain an understanding of God. For whatever system the human mind may occupy, there are properties that cannot be fully understood within that limited system. Since God occupies all higher systems as well, the limited human mind can never reach these higher levels and comprehend the divine.

For mathematicians, Gödel's theorem held even greater

prominence than it did for philosophers. What Gödel has taught us is that some theorems can never be proven. This is a very disturbing idea for many reasons. The aim of mathematics is to build a structure of truths: theorems and lemmas and corollaries, all constructed step by step from a basic set of principles called axioms, using the laws of logic. Gödel's proof of the incompleteness theorem demonstrated that no matter how careful mathematicians may be in designing a logical system of first principles on which to construct arithmetic, algebra, analysis, and the rest of mathematics, such a system can never be complete. In any such system, there will be questions that cannot be answered. The system will always contain issues that are undecidable, regardless of whether or not they are true.

While Gödel did not become an international celebrity overnight, many were quick to understand the immeasurable significance of the results he proved when he was only twenty-six years old. Gödel made an announcement of his results at a seminar at the University of Vienna the day after he presented his dissertation in 1930. Within a year, prominent mathematicians on both sides of the Atlantic were familiar with the important theorems Gödel had proved, and he was invited to spend a semester at the newly founded Institute for Advanced Study at Princeton. By then Gödel was working as a *Privatdozent*, as Cantor had done half a century earlier, eking out a living by privately tutoring students at the university. Even so, he was reluctant to leave his beloved Vienna for America and the promise of much higher pay. He was also reluctant to leave Adele behind. In the fall of 1933, however, Gödel finally embarked on his first trip to Princeton to take up a fellowship at the Institute. There he would meet Einstein for the first time and start a lifelong friendship.

Once Gödel had shown that some results in mathematics could never be proven, he turned his interest to a result that by then could not be proved: the continuum hypothesis. Another issue that interested him was the problematic axiom of set theory: the axiom of choice. Gödel began to address these two problems.

But soon after he touched the forbidden concept of the alephs and actual infinity, Gödel—like Cantor a few decades earlier—began to exhibit symptoms of mental illness. Gödel, who had never been harassed by colleagues and had no mortal enemies of the caliber of Kronecker, began to show signs of mental imbalance strikingly similar to those of Cantor. He became depressed and slowly developed an obsessive conviction that people were persecuting him. He became suspicious that someone was trying to poison him and, within a few years, would demand that Adele—by then his wife—taste all food before he touched it. As the disease progressed, Gödel ate less and less. Within almost half a century, he starved himself to death.[38]

In 1934, Gödel was again invited to stay at the Institute for Advanced Study, where great mathematical minds appreciated his abilities and genius. The world was changing fast as the storm clouds of Nazism were gathering over Europe. As was the case during World War I, Gödel was oblivious to the dangers and hatreds surrounding him—even though he lived in Vienna and could not have been unaware of the Nazis' rise to power in Germany. Antisemitism was rampant in Vienna, and many of his Jewish friends, including his adviser Hans Hahn, were already being harassed by mobs and by the university administration. Gödel himself was attacked by a gang of Nazi thugs on a Vienna street when he was taken for a Jew because he was dressed in black.[39]

Anyone would have jumped at the opportunity to leave Europe for a chance to live in America when Europe was about to burn. But Gödel was oblivious to his surroundings. His work in the foundations of mathematics was absorbing him, and, slowly, he was losing contact with reality. He suffered from strange ailments—stomach pains, breathing problems, and other difficulties that can be either real or psychosomatic. In the summer, his illness became worse and he checked himself into a sanatorium near Vienna. This was a lavish institution designed to care for the mental health of wealthy patients. Here, Gödel spent several months regaining his health. He wrote to the director of the IAS, asking to postpone his trip to America. He did not mention his hospitalization.

After his release from the hospital, Gödel returned to Vienna and taught a summer course at the university. Throughout this period, he was working feverishly, trying to prove that the axiom of choice was consistent with the rest of the axioms of set theory.

In September 1935, Gödel left again for America, sailing from Le Havre on board the *Georgic*, whose passengers included the famous physicist Wolfgang Pauli and the mathematician Paul Bernays, both of them also headed for the Institute for Advanced Study.[40]

But in November, Gödel resigned his post at the Institute. He was suffering a relapse of depression and wanted to return to Vienna to recuperate under the care of his fiancee. At the height of his depression, however, Gödel found a proof of the consistency of the axiom of choice with the rest of the foundations of mathematics. Next he tackled the greatest problem of all: the mystery of the alephs and the continuum hypothesis. On December 7, 1935, Kurt Gödel arrived at Le Havre and continued on to Paris. He called his brother in Vienna, asking him to come and escort him home.

ℵ18

The Night of June 14–15, 1937

Gödel was content to go back to a Europe about to erupt in the worst conflict in history, and he remained oblivious to the portentous events of the late 1930s. In Vienna, he and Adele were making plans to marry, which they would do after Germany's *Anschluss*—the annexation of Austria in 1938—in yet another indication of how utterly insulated their life had become from events around them.

Upon his return to Vienna, Gödel is said to have made the statement: "Jetzt, Mengenlehre!" ("Now, set theory!"), indicating his desire to devote all his energies to working on the impossible problems of Cantor's actual infinity. He must have been aware that his concentration on these topics was slowly driving him mad, but, like Cantor, he was drawn to the infinite light like a moth to fire. In and out of sanatoria, the world around him crumbling, Gödel ignored the generous invitations from America—which now included not only the IAS but other universities as well—and immersed himself in the study of the continuum hypothesis.

His attempts to solve the mathematical problem exacerbated his mental problems. Gödel became convinced that he was being poisoned by the "bad air" he was breathing. This air came from the refrigerator in the apartment he and Adele now shared, and from the heating system. Adele was insensitive to his problems—she smoked throughout her life, even though she knew her cigarettes were adding to the "bad air," real or perceived. Throughout 1937 and 1938, Gödel was largely incapacitated. The couple was living off their savings, which were fast becoming depleted. Gödel's only income came from a course on axiomatic set theory he was teaching at the university.

Gödel kept copious notes he wrote to himself in shorthand. Years later, after he had published an important paper proving the consistency of the continuum hypothesis with set theory, his original notes surfaced. A cryptic note he wrote in a notebook reads: "Kont. Hyp. im wesentlichen gefunden in der Nacht zum 14 und 15 Juni 1937." (Continuum hypothesis found significant in the night of 14 and 15 June 1937.) Gödel found a way to prove that the continuum hypothesis was consistent with the axioms of set theory. Whether or not Cantor's hypothesis about infinity was true, assuming that it was true did not create any new contradictions within the foundations of mathematics.

Surprisingly, Gödel told no one about his important discovery. In 1938, he returned to Princeton for a semester at the IAS, and then went on to teach for a semester at the University of Notre Dame. People who knew him while he was at these institutions described him as morose and brooding. Undoubtedly, he was depressed and lonely, separated as he was from his wife, who remained in Vienna. When he returned home, Austria was part of Germany and World War II was beginning. In 1939, Nazi authorities in Vienna found Gödel fit for service in the army of the Third Reich.

The Nazi order was perhaps the final straw. Gödel was at last made to face reality—he was on a continent at war, and unless he did something about it, he would have to take an active part in the violence. Since he was opposed to the Nazis, it was time to leave. Gödel used his standing invitation to the IAS to try to obtain a visa for the United States. But his move came almost too late. Everyone who could apply to escape from Europe was doing so, and the U.S. Embassy in Vienna was not making the process easy for refugees.

Meanwhile, Gödel's paper on the consistency of the axiom of choice and of the generalized continuum hypothesis with the rest of the axioms at the foundations of mathematics appeared in the *Proceedings* of the National Academy of Sciences of the United States. The paper combined Gödel's two great achievements after the incompleteness theorem. He had proved that the axiom of choice and Cantor's hypothesis as generalized by Hausdorff were *consistent*. Both enigmatic statements about infinity, if assumed true, did not create any contradictions or clash with the rest of the axioms of set theory. If the foundations of mathematics were sound, Gödel's proof implied, they would remain sound if we assume the truth of both statements. His proof did not imply that the continuum hypothesis (or the axiom of choice) was true. In fact, Gödel's result was halfway toward proving that the continuum hypothesis (and the axiom of choice) was *independent* of the rest of mathematics. While struggling to obtain a visa to America and survive in the hostile environment of wartime Vienna, Gödel was working to complete the proof of the reverse proposition and thus establish independence.

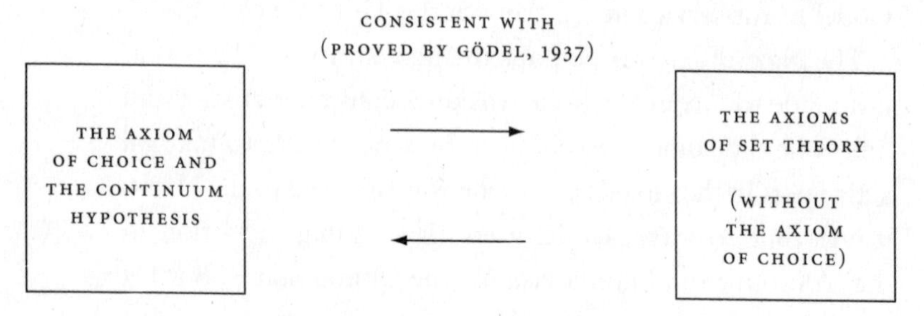

Gödel's proof, completed during the night of June 14-15, 1937, implied that the continuum hypothesis might work within the set of axioms forming the foundations of mathematics. If the back-implication was proved, the axiom of choice and the continuum hypothesis would be shown to be completely independent of mathematics, meaning that within the present system we can't know whether or not Cantor was right about the orders of infinity.

With Gödel's enhanced fame (at least within international mathematical circles), the director of the IAS at Princeton could use his influence to demand a visa for the couple. Nonetheless, the Gödels had trouble both with the embassy and with the Nazi authorities ruling Austria, who were suspicious of an Austrian professor's bid to immigrate to the United States. Ultimately, the difficulties were overcome and a visa was issued, followed by a permit from the authorities to leave Europe. But the Gödels were a little late—the usual route from Europe to America across the Atlantic was now closed because of the war. But a solution was found—the Gödels would travel across Siberia to the Pacific coast of Asia, then continue to Japan, and from there they would sail to San Francisco.

It was now 1940, and Europe was at war. Within a year, Pearl Harbor would be attacked and America would join the World War. It was already dangerous to travel across Europe and Asia. The Gödels boarded a train and traveled the Trans-Siberian Railway. At many stops along the way they were almost sent back, but managed to make it safely to Japan. On February 20, they boarded the *President Cleveland* bound for San Francisco. When they arrived at Princeton a few weeks later, Gödel expressed surprise that there were so many refugees trying to flee Europe. Even after the difficulties he and his wife had experienced in wartime Europe, he was living in a strange internal world. When asked what life was like in Vienna before they left, Gödel replied: "The coffee was wretched."[41]

$\aleph_{19}$

Leibniz, Relativity, and the U.S. Constitution

Gödel, like Cantor, couldn't handle the intensity of actual infinity for long. He was descending into madness from the intense introspection that studying the continuum hypothesis entailed. He knew that the continuum hypothesis did not cause contradictions when placed within the system containing all the axioms of set theory. He had also proved that the contentious axiom of choice was "safe" to use within these other axioms. But he did not know whether the converse was true. He did not know whether the negation of the continuum hypothesis and the negation of the axiom of choice were also consistent with set theory. He spent many months brooding sullenly, trying to prove that the continuum hypothesis was completely independent of the rest of mathematics. This would be the case if the negation of the continuum hypothesis was also consistent with the axioms of set theory.

Gödel spent the summer vacationing on the coast of Maine. He whiled away the hours at night walking by the woods on the beach in deep concentration. People at the resort knew him as "German," and when they saw him walking on the beach alone at night, some thought he was a spy waiting to send secret signals to a U-Boat. Gödel continued to lose touch with reality. The flowers at the resort made him think he was at Marienbad.

Gödel was becoming more paranoid. He was convinced that his doctors were trying to have him committed. Every heating or air conditioning system was spewing poison gases. His food was poisoned. He would send for oranges, but when they arrived he would say they were no good and send them back. He tried to devise a mathematical proof of the existence of God. He thought he had a proof, then decided it wasn't good, then again he had it, then not. He was still working on the continuum problem, but finally—like Cantor before him—gave up all attempts to solve the problem and became obsessed with other issues. While Cantor spent years on his futile attempts to prove that Shakespeare didn't write his plays, Gödel now spent years trying to prove that Leibniz developed theories that were probably not his. Like Cantor, Gödel found no convincing proof of his assertions. There was something otherworldly about the continuum hypothesis, something that made it impossible to contemplate for long. Trying to prove it was hazardous to the mind; the alternative was to abandon it and move on to a different field.

After the war, Gödel made contact with German libraries in an effort to obtain photocopies of all of Leibniz's works. These papers were numbered in the tens of thousands and were already under study by a number of other scholars. After wasting many years, like Cantor with his Shakespeare obsession, Gödel gave up. In the meantime, he developed a warm, close relationship with another genius escapee from Europe working at the IAS—Albert Einstein.

Many people were puzzled by the fact that Einstein and Gödel became good friends. The two men's personalities were very different. Einstein was open, gregarious, and had a sense of humor. Gödel was closed, staid, and had few friends. When asked what made Einstein enjoy his company, Gödel replied that the reason was that he never agreed with Einstein, and that their arguments were what Einstein found attractive. An obvious reason for the friendship was also that Einstein preferred to speak German, and Gödel was a native speaker of that language. Gödel and Einstein spent many hours talking about physics and philosophy and mathematics. Through their conversations, Gödel became increasingly interested in the theory of relativity. Here was another escape for him from infinity and the continuum hypothesis. Relativity was a challenging field in which his genius could be put to use, but without the unbearable light of the infinite driving him mad.

So Gödel began to work on relativity theory on his own. He started with Einstein's field equation of gravitation, applied to the universe as a whole. Gödel attempted to solve that equation to find what kind of solutions he could obtain and what these solutions might say about the universe and space and time. The results were surprising. Gödel assumed a universe that rotates, does not expand, and is homogeneous—a universe that looks the same everywhere. His solution of Einstein's equation within this framework implied that time travel was possible. Today we know that the universe is expanding and probably not rotating, so Gödel's solution is not valid for this universe. However, when his paper came out in 1949, it received much attention. Gödel thus made an important contribution to the field developed by his friend Einstein. The two continued their discussions with more excitement since they now shared some important contributions to science.

About the same time that his friendship with Einstein was developing, Gödel applied for American citizenship. It is not clear why he qualified to apply, since U.S. immigration laws exclude persons with mental problems, especially those with histories of hospitalization. Nevertheless, Gödel was scheduled for his citizenship test. In preparation for the test, he had to study U.S. history and civics. He studied the U.S. Constitution. But Gödel read the document the way only a logician could. He studied every sentence with incredible care, looking for logical lapses or paradoxes. And he claimed to have found some. One day, not long before his scheduled hearing, Gödel rushed into Einstein's office and exclaimed: "I've found a logical inconsistency in the Constitution!" Einstein was understandably alarmed, since he realized that if Gödel made such statements at his immigration hearing, his case would be imperiled. He enlisted some of the other members of the IAS in an effort to assuage the excited Gödel.

But the efforts failed. Gödel was not to be talked out of his great discovery, and he expounded on the Constitution to anyone who would listen to him. The day arrived and Gödel had to appear before the judge deciding his immigration case. In the car, Einstein and another friend did their best to distract Gödel from his "discovery," but to no avail. The hearing began, and Gödel wasted no time in telling the judge that the Constitution was flawed and could allow a dictatorship to establish itself in the United States, as had happened in Europe. Fortunately, the judge had both patience and a sense of humor—and he was flattered to have the famous Einstein in his courtroom as a witness for the applicant. Gödel was granted his citizenship.

After these diversions into other areas, Gödel did return briefly to consider the continuum problem. But he never again published on it. In fact, after 1958, he published nothing at all. Gödel's thoughts about infinity were now contrary to Cantor's original ideas. Gödel did not believe that the continuum hypothesis was true. Over the years, he kept changing his mind about the power of the continuum. First he thought that c was $\aleph_2$, then he decided that it was "another aleph," and afterwards his mind oscillated among other alternatives, none of them $\aleph_1$. He seemed to resign himself to the fact that he would never make progress on the problem. But in 1963, new light was shed on the continuum hypothesis and on the axiom of choice from an unexpected source.

$\aleph_{20}$

Cohen's Proof and the Future of Set Theory

In the spring of 1963, while the 57-year-old Gödel was preoccupied with his health, his paranoia, and his notions about the continuum and infinity—no longer able to commit to the intense concentration required for such work—an important development was taking place on the other side of the continent.

Paul Cohen, a young mathematician at Stanford University, used an ingenious new method called "forcing" to prove that the axiom of choice was independent of the other axioms of set theory, and that the continuum hypothesis was indeed independent of all the axioms put together, including the axiom of choice. Cohen accomplished this task by proving the complement of the result established years earlier by Gödel. Cohen's proof showed definitively that the truth of Cantor's continuum hypothesis could not be established within the current system of axioms of set theory.

This did not necessarily mean that Gödel's incompleteness result applied to the continuum hypothesis—it was still possible that another system of axioms might allow for a proof of either the truth of the continuum hypothesis or of its negation. What Cohen's proof told us was that within the current system of axioms, the continuum hypothesis could be taken either as true or as false and no new contradictions would result. So after all the years of hard work in an effort to find out whether Cantor was right or wrong, the continuum hypothesis remained an enigma.

To prove the continuum hypothesis or to prove that it was wrong (and thus that there were other alephs between aleph-zero and the power of the continuum) would require a different axiom system. But which axiom system could be used? Zermelo-Fraenkel was the best system, one that had served mathematics well; how could logicians find a system to replace it? Any other system was likely to include inconsistencies or errors. Zermelo-Fraenkel had survived the test of time and held many important properties of axiom schemes. But it could not tell us in any way whether or not the continuum hypothesis was true.

Paul Cohen received his Ph.D. in mathematics from the University of Chicago in 1958. At the time, he was working in harmonic analysis—an area far from the foundations of mathematics—and showed no interest in logic or foundations. Cohen went on to solve a very important problem in Banach algebras, and his new renown in mathematics earned him an invitation to spend the years 1959–61 at the IAS at Princeton. Surprisingly, there is no evidence that he ever met Gödel during the time they were both at the IAS.

Cohen wanted to solve another great problem in mathematics, and while at the Institute, he befriended the logician Solomon Feferman and asked him about problems in logic and foundations. After his stay at the IAS, Cohen took a position at Stanford University and continued to work on foundation problems there. At one point he decided that the most important problem in logic and foundations—in fact, in all of mathematics—was a proof of the independence of the axiom of choice and the continuum hypothesis from the rest of the axioms of set theory. Cohen knew that Gödel had proved one direction of the independence already, and that a proof of the other direction was now needed.

Cohen consulted with a number of logicians at Stanford, and worked hard on the problem. Then he devised his new proof technique, forcing. Using a clever argument, Cohen was able to force a set of postulates to take one of two values. The forcing method entailed starting with a collection of sets and rules of logic that apply to these sets, and then enlarging the system gradually so that the rules still applied to an ever-larger collection of sets. Manipulating the postulates within the larger logical system gave Cohen his final answer. The continuum hypothesis was now proved to be completely independent of the axioms of set theory. Whether true or not, it would be mathematically impossible to prove the hypothesis or to disprove it within the current system. Along with the continuum hypothesis, the axiom of choice was also shown to be independent of the rest of the axioms of set theory.

When Cohen's proof was scrutinized by experts, some errors were discovered, but he corrected all of them. Still, Cohen was worried about his proof and decided to send it to Gödel to see what he thought about it. Gödel read Cohen's proof and saw that it was indeed a great mathematical result—one that he himself had been trying unsuccessfully to obtain over many years of work. He hailed Cohen's important result and encouraged him to publish it. In 1966, Cohen received the Fields Medal—the greatest honor in mathematics—for his achievement. This was the first and only time that the Fields Medal was awarded for work in the area of logic and foundations.

Like Gödel in his later years, Cohen doubted the actual truth of Cantor's hypothesis. In a lecture at Harvard University in the 1960s, described by Dauben, Cohen expressed his opinions on the continuum hypothesis.[42] According to Cohen, the continuum is such a rich set that it is unlikely that its power (its cardinal number) is aleph-one. Lower cardinals are accessible from one another by certain mathematical operations. But according to Cohen the continuum lies far above the lower infinities—its cardinal number is much larger than $\aleph_1$.

The problem is of course that results in mathematics require proof. When a mathematician—gifted and renowned as he or she may be—makes a statement about infinity or the continuum, the statement has to be proved. Unproven statements carry little weight in the world of mathematics. Gödel and Cohen have shown us that a proof of the continuum hypothesis is impossible within the current system; so until such a time when we are able to construct another system, the continuum hypothesis will remain an enigma.

Mathematics had to adjust to the reality that a major hypothesis within its very foundation could not be proved. As a result, mathematicians have proposed a wide variety of alternative hypotheses. Unfortunately, these other assertions about the continuum also remain unproven and undetermined.

The weak continuum hypothesis, suggested by F. B. Jones, is the assertion:

$$2^{\aleph_0} < 2^{\aleph_1}$$

Cantor had shown that the cardinality of a set is always smaller than the cardinality of its power set; therefore, $\aleph_1 < 2^{\aleph_1}$. Thus, *if* the continuum hypothesis is true, that is, the statement $2^{\aleph_0} = \aleph_1$ holds, then the weak continuum hypothesis is also true: $2^{\aleph_0} < 2^{\aleph_1}$. (To see this, simply substitute $2^{\aleph_0}$ for $\aleph_1$ in the equation.) Since we don't know if the continuum hypothesis is true, there is no way of telling whether the weak continuum hypothesis is true. The negation of the weak continuum hypothesis is the statement that $2^{\aleph_0}$ is *equal to* $2^{\aleph_1}$. This assertion is the Luzin hypothesis, named after Nikolai N. Luzin, a Russian mathematician who began to study Cantor's continuum problem in 1916. Luzin worked in Moscow and had a number of talented students, who together with him formed the Moscow school of function theory. These Russian mathematicians were interested in the continuum hypothesis and its implications in mathematical analysis and other areas. They proposed and developed important results.

L. Bukovsky has shown that the Luzin hypothesis is consistent with set theory. Therefore, the weak continuum hypothesis (the negation of the Luzin hypothesis) is independent of set theory. As a consequence, all three statements remain a mystery.

Many results in mathematics, such as topological properties of various spaces, cannot be determined without knowing whether the continuum hypothesis, its weaker form described above, or its converse—the Luzin hypothesis—is true. Therefore, mathematicians must at times insert a statement within the proof of a theorem stating that the result depends on the continuum hypothesis or on a related unproved result. The completion of the theory of mathematics will have to wait until the day when more is known about the mystery of the continuum.

The lack of knowledge, however, has not stopped the development of set theory beyond the foundations laid by Georg Cantor and his successors Zermelo, Gödel, and others. Kurt Gödel made the first attempts to advance set theory and the rest of mathematics forward. Gödel suggested the possibility that *very large* numbers—infinite cardinals that are even larger than the "usual" infinite cardinals—might exist. This notion gave rise to the modern area within set theory: large cardinals. Large cardinals have not been proven to exist. These exceptionally large infinite quantities are made to exist by decree: set theorists have developed a collection of axioms that establish the potential for existence of these new cardinals. The idea behind large cardinals is an abstraction of the difference between a finite number and the first infinite cardinal, $\aleph_0$. The argument proceeds as follows.

We cannot reach the first infinite cardinal by any mathematical operation (addition, multiplication, or exponentiation) if we start with a finite number such as five, or five hundred trillion, for that matter. The first infinite cardinal, $\aleph_0$, is thus unreachable, or *inaccessible* from any of the finite cardinals (any given number is a finite cardinal). Once we have reached the lowest infinite cardinal, $\aleph_0$, however, we can obtain higher infinite cardinals by exponentiation, since by Cantor's theorem the power set of any given set has higher cardinality, and thus $2^{\aleph_i}$ is a higher cardinal number obtained from $\aleph_i$. This is true even if the continuum hypothesis is false (in which case there might be some infinite cardinal *between* $\aleph_0$, and $2^{\aleph_0}$).

The idea of large cardinals is that perhaps $\aleph_0$ should not be the only infinite cardinal that enjoys the "inaccessibility" quality. If this assumption is correct, then somewhere in the vastness of the infinite field of infinite cardinals, there exist other infinite cardinal numbers that are unreachable from lower infinite cardinals. Such cardinals must be so huge that they are not reachable from any lower infinite cardinal by exponentiation or by any other mathematical operation performed on a lower infinite cardinal. This is what the universe of cardinals looks like, if large cardinals do exist.

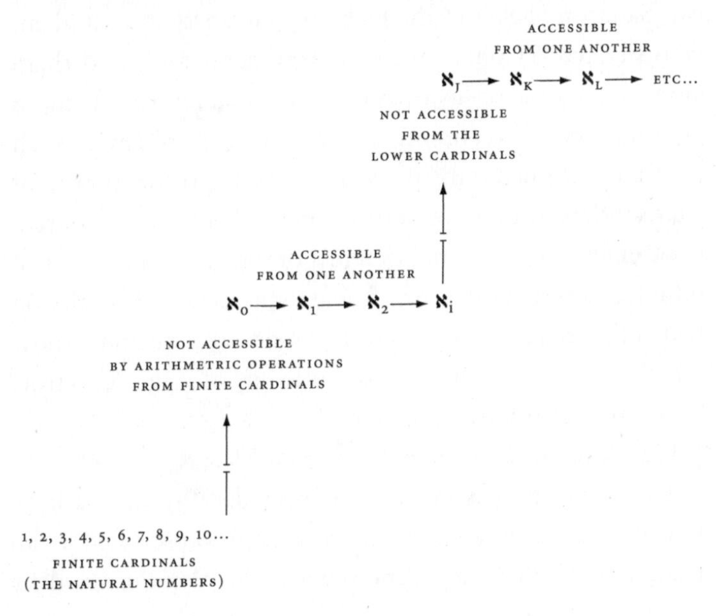

Set theorists have worked extensively with these gargantuan infinite numbers: infinities that dwarf all other infinite quantities. Consequently, research on large cardinals has produced many interesting and important results in set theory. Some of these new findings have shed light on what happens at lower cardinal levels, although, so far, large cardinals have not brought us a solution to the continuum problem itself.

The torch of set theory has been passed on once again. From Cantor it went to Zermelo, then to Gödel, and from him to Cohen. Following Cohen's invention of the forcing method, set theory reached a dead end in the 1960s. But in 1974, Jack Silver of the University of California at Berkeley proved an important result about the alephs, which opened the way to new research by young set theorists. Among them was Saharon Shelah of the Hebrew University in Jerusalem, who perfected Cohen's forcing arguments and used them along with some of his own methods of analysis to prove a large number of theorems about infinity. Here is one such theorem, included only in order to give the reader a taste of some of the newest developments in set theory. This theorem exemplifies the depth and richness of the modern theories of infinity, as well as the inexorable dependence of results on truths that remain beyond our reach because we don't know whether the continuum hypothesis in any of its guises is true.

Shelah's Theorem:

If $2^{\aleph_n} < \aleph_\omega$ for every number n, then $2^{\aleph_\omega} < \aleph_{\omega_4}$.

The subscript 4 is indeed unexpected. Why should it be that if 2 raised to the power of any aleph subscripted by an integer is smaller than aleph-omega (the aleph indexed by the first transfinite ordinal), then two raised to the power aleph-omega must be less than the aleph indexed by the *fourth* uncountable number? We don't know, and few people in the world have any intuition about such complex levels of infinity. But Shelah indeed proved this theorem.

ℵ21

The Infinite Brightness of the Chaluk

Gödel and Cohen have brought us to a sobering realization: hard as we may try, there will always be some truths forever beyond our reach. Human beings may never understand the deep nature of infinity. This is perhaps something that the Kabbalah practitioners understood on an intuitive level, without requiring a mathematical proof. To them, infinity was God or things that are God's. One such infinity was the *chaluk*, God's infinitely bright robe, at which no human could look.

But a handful of people in history have been given a glimpse of infinity. The keen minds of ancient Greece, just as human civilization was awakening, were able to grasp surprisingly abstract truths about infinity—as Zeno's paradoxes and the work of Archimedes, Eudoxus, and others can attest.

Galileo, the Renaissance father of physics with his uncanny sense for the workings of the universe, was blessed towards the end of his career with a fleeting look at a property of discrete infinity. Bolzano, a cleric mathematician, was able to leap to continuous infinity and to understand the paradoxical nature of infinite sets on the real line.

But it was Georg Cantor, the lone creator of modern set theory, who truly understood some important truths about infinity, and could separate the concept into different levels. Trying to understand the real meaning of the various levels of infinity—trying to dissect the unreachable infinite and probe its innermost parts—may have cost him his sanity. But Cantor's work opened a door to paradise, one that could never be closed again. For mathematics after Cantor would not be the same, whether because of the properties of the infinities he had discovered, or because of the terrible paradoxes and pitfalls he and his contemporaries had exposed. With infinity understood on some level, and with the dangers of venturing further into its web more apparent than ever, mathematics has matured in the last century into a more coherent and better-organized discipline.

Hand in hand with mathematics, computer science has emerged as an immensely important area in the modern world. And here, too, infinity and its study—and the limitations that beset us when we try to understand the nature of infinity—have left their mark.

In 1936, Alan Turing showed that no mechanical procedure could solve the "halting problem." The halting problem is the question of whether a given computer program will eventually stop. A real number is computable if there is a computer program for calculating its digits one by one. Surprisingly, almost all real numbers are not computable. Turing proved that if we could find a mechanical procedure to decide whether a given computer program will eventually halt, then we could compute a real number that is not computable, which is a contradiction. The problem, and the research in computer science that has flourished over the decades since Turing's proof, is related to Gödel's work. With all their dizzying power, computers, like people, are still stymied by infinity.

In physics, the idea of infinity materializes when we consider the extent of the universe: Is the universe finite or infinite? The answer to this question is of course not known. Neither do physicists know whether actual physical space can be infinitesimally divided. Some theories posit the existence of a "smallest measure" of space and time, related to the Planck time—a basic unit of measure. In string theory, there is an assumption about the existence of a smallest element, a tiny string, which cannot be subdivided. But physicists have no proof that such entities do indeed exist. We are therefore left with the open question of whether infinity has a meaning in the physical world.

We know that numbers extend infinitely, and Cantor has taught us that such infinities are hierarchical in nature: one larger than the other to an infinite extent. But here a key question arises:

Do numbers actually exist?

It would seem that numbers are merely abstractions that people have proposed to facilitate counting and comparison of actual, physical quantities. Some say, therefore, that numbers form a language or a convention invented by people in order to address problems of the real world. But in this case, we should know everything about numbers and their interrelations, since we have invented them. And this, I believe, is not the case. In mathematics we constantly discover the properties of numbers (and of functions and spaces and abstractions of numbers) through hard work—often finding truths that go against what our intuition tells us. Numbers, therefore, cannot be our invention. They are entities about which we learn fascinating new things all the time. The study of numbers and their abstractions and associated concepts is what mathematics is all about.

Numbers do exist, and their existence, I believe, is independent of people. In another universe, one without people and without anything we recognize from our own universe, numbers will still exist. And these numbers are infinite. But how densely packed are these numbers?

Does the continuum really exist?

Is it possible that, while numbers do exist, the continuum does not? This notion is reflected in Leopold Kronecker's statement: "God made the integers and all the rest is the work of man." The question of the existence of the continuum is haunting. We have no evidence from the physical world that anything can be subdivided ad infinitum. However, the field of calculus, which makes essential use of the assumption of infinite divisibility and the existence of the continuum, works astonishingly well in giving us precise answers to real world problems. Why would such a continuum-based approach be so effective if the continuum itself did not exist?

א

A key tenet in the Kabbalah is that the Ein Sof contains the

Ein. The infinity of God also contains nothingness. For there is nothing within things that are of God beside infinity. In mathematics, too, infinity contains emptiness. An infinite set also contains the empty set. This statement can be understood within the context of the foundations of mathematics. Peano defined numbers starting with nothingness. He defined zero as the empty set—pure nothingness. He then defined the number one as the set containing the empty set; the number two as the set containing both the empty set and the set containing the empty set; and so on. Peano thus defined the infinitude of numbers starting with nothing at all.

The road to understanding infinity may never lead us to a complete understanding of its deepest properties—some knowledge may never be ours. It seems likely, however, that research in mathematics over the coming decades will give us important results in this direction. As often happens in mathematics, a research effort aimed at discovering a particular truth inevitably leads elsewhere and teaches us new things. We may some day be able to develop a more consistent foundation for mathematics, freer of paradoxes and other difficulties. Within such a new system, the continuum hypothesis might be seen in new light, perhaps illuminating the nature of the alephs and where the power of the continuum lies within them.

So while we may never be able to withstand the intense light of the chaluk, we may be able to learn about smaller lights along the way to human understanding and knowledge.

א

Georg Cantor believed that the two questions above are both answered in the affirmative. In an 1883 paper, Cantor said that numbers have an *intrasubjective and immanent reality*.[43] Cantor firmly believed that higher infinities, his alephs, exist as well, and that the continuum is also real. According to Cantor, physical reality does depend on mathematical principles. Thus, the continuum, numbers, and their properties are all reflected in various aspects of physical reality. But mathematics, Cantor added, does not need the physical world to justify its existence. Mathematics, and the infinite levels of infinity—the transfinite numbers—have a meaning all their own.

א

A short distance outside the historical center of Halle there is a large residential area built during Soviet times. Here, wedged among several large buildings, each containing hundreds of apartments, is a small grassy area where local residents walk or relax. At its center, a commemorative bronze plaque was placed in 1970 on a cement block. From a distance, the plaque looks like a memorial to Lenin. But the brass relief is of Georg Cantor. Next to his face is an array of numbers with arrows leading from one number to another, symbolizing the diagonalization proof. Below it is a mathematical equation related to the continuum hypothesis. Underneath all the symbols is the sentence that probably captures Cantor's deepest conviction about mathematics. Translated into English, it says: "The essence of mathematics lies in its freedom."

Appendix

The Axioms of Set Theory

The following axioms of set theory are the ones generally in use in mathematics. They are intended to form a basic structure over which the foundations of mathematics are built. The axioms were designed by Ernst Zermelo and other logicians at the beginning of the twentieth century. After years of searching, the mathematicians found a minimal set of assumptions that lead to a consistent body of knowledge including the natural numbers, the real numbers, complex numbers, and their properties and arithmetic. Other areas of mathematics such as geometry, topology, algebra, also follow from this base upwards to form the edifice of the entire field of mathematics. The axioms were largely inspired by the work of Georg Cantor, who had implicitly assumed some of these axioms in his founding of the theory of sets.

1. The Axiom of Existence:

There exists at least one set.

The one set in this axiom can be taken as the empty set. Other sets are then constructed from it. One of them is the set containing the empty set, and so on.

2. The Axiom of Extension:

Two sets are equal if and only if they have the same elements.

3. The Axiom of Specification:

To every set A and every condition $S(x)$ there corresponds a set B whose elements are exactly those elements x of A for which $S(x)$ holds.

This is the axiom that leads to Russell's paradox. For if we let the condition $S(x)$ be: not(x element of x), then $B=\{x$ in A such that x is not in $x\}$. Is B a member of B? If it is, then it isn't; and if it isn't, then it is. Therefore B cannot be in A, meaning that nothing contains everything.

4. The Axiom of Pairing:

For any two sets there exists a set to which they both belong.

5. The Axiom of Unions:

For every collection of sets there exists a set that contains all the elements that belong to at least one of the sets in the collection.

6. The Axiom of Powers:

For each set there exists a collection of sets that contains among its elements all the subsets of the given set.

Cantor has shown that the power set is always larger than the set itself. This led to the paradoxical conclusion that there cannot be a largest set, or a largest cardinal number. For whatever set you take as the "largest," there always exists the power set of this set, which is larger—leading also to a larger cardinal number than that of the original set you designate as largest.

7. The Axiom of Infinity:

There exists a set containing 0 and containing the successor of each of its elements.

This axiom aids in defining the natural numbers. We start with zero, adding one to get the first natural number, 1, then adding one to get two, and so on to infinity.

8. The Axiom of Choice:

For every set A there is a choice function, f, such that for any non-empty subset B of A, $f(B)$ is a member of B.

The choice function assigns ("chooses") a member from each set B. The problem with the axiom of choice lies in the fact that there may be infinitely many sets B within A.

א

Author's Note

The idea for this book began to germinate in my head late one night twenty-five years ago, while talking to my friend Bob Trent, who was a graduate student in mathematics at the University of California at Berkeley. We were both beyond tiredness and surviving on multiple cups of coffee when Bob said: "Look, I want to show you something," and proceeded to write down a sequence of symbols: 1, 2, 3, . . . , $\omega+1$, $\omega+2$, . . . 2ω, . . . ω^2, . . . ω^ω, ... I was fascinated by the idea that the natural numbers could be continued beyond infinity, and that we could actually talk about different levels of infinity, getting larger and larger with no end. I was enthralled by the paradoxes Bob explained to me about the impossibility of a largest infinity or a set containing all sets. This was the heart of mathematics, I knew.

Then I learned the story of the tormented life of the man who first proposed the ideas of actual infinity and the continuum hypothesis. I was captivated by what I learned about the life of Georg Cantor. Years later, when I told the story to my publisher, John Oakes, he suggested that I write a book about it. I am grateful to John for encouraging me to pursue this story over the last five years, and for his patience and unwavering support throughout the years it took me to research and write this book. I am grateful to the dedicated staff at Four Walls Eight Windows for all their hard work in producing this book: JillEllyn Riley, Kathryn Belden, and Philip Jauch.

I am grateful to Professor Daniel Ruberman, Chair of the Mathematics Department at Brandeis University, for arranging for me to spend the year as a Visiting Scholar in Mathematics while I was writing this book. I am grateful to the librarians at Brandeis University and at Bentley College for ordering for me many obscure documents, articles, and books relevant to the work of Georg Cantor and the concept of infinity.

I wish to express my deep gratitude to the mathematicians who were generous with their time and were willing to be interviewed for the purpose of preparing this book. They include Professor Akihiro Kanamori of Boston University, Professor John Dawson of the Pennsylvania State University, and Professor Saharon Shelah of the Hebrew University in Jerusalem.

I thank Professor Manfred Goebel and Professor Karin Richter of the Mathematics Department at the Martin-Luther University of Halle, Germany, for their hospitality during my stay in Halle, and for guiding me to many papers, photographs, documents, and sites relevant to the story of Georg Cantor, his life and his mathematics. I am also grateful to the

Georg Cantor Society at the University of Halle for its hospitality during my stay at the University.

I thank Dr. Frank Pillmann, a psychiatrist and administrator at the Halle Nervenklinik, for sharing with me his thoughts about Georg Cantor's illness and for showing me the clinic and the areas in which Cantor was hospitalized during the first few years of the twentieth century. I am also grateful to Dr. Pillmann for showing me various documents describing the clinic and its history, as well as Cantor's records of hospitalization.

I am grateful to the mathematicians Dr. Eugene Pinsky and Dr. Yakov Karpishpan for various discussions about the mathematics of Georg Cantor, and to Professor Underwood Dudley of DePauw University and Dr. Don Albers, Director of Publications of the Math Association of America, for their many comments on the entire manuscript. Their contributions improved the text. I am grateful to psychiatrist Dr. Venyamin Pinsky and psychologist Dr. Idell Goldenberg for discussions of mental illness.

Finally, I am grateful to my wife, Debra, for her help, support, and encouragement throughout the preparation of the manuscript.

References

Afterman, Allen. *Kabbalah and Consciousness.* Riverdale, NY: Sheep Meadow Press, 1992.

St. Augustine. *City of God.* New York: Penguin, 1972.

Barrow, John D. *Pi in the Sky: Counting, Thinking, and Being.* New York: Oxford University Press, 1992.

Bell, E. T. *Men of Mathematics.* New York: Simon & Schuster, 1937.

Bell, John. *Boolean-Valued Models and Independence Proofs in Set Theory.* NY: Oxford University Press, 1985.

Benacerraf, P. and H. Putnam, eds. *Philosophy of Mathematics.* Englewood Cliffs, NJ: Prentice-Hall, 1964.

Bolzano, Bernard. *Paradoxes of the Infinite.* New Haven, Conn.: Yale University Press, 1950.

Boyer, Carl B. *A History of Mathematics.* New York: Wiley, 1968.

Boyer, Carl B. *The History of The Calculus and Its Conceptual Development.* New York: Dover, 1959.

Bunch, Bryan. *Mathematical Fallacies and Paradoxes.* New York: Dover, 1982.

Cantor, Georg. *Contributions to the Founding of the Theory of Transfinite Numbers.* Translated by Philip E. B. Jourdain. La Salle, IL: Open Court, 1952.

Charraud, Nathalie. *Infini et Inconscient: Essai sur Georg Cantor* (in French). Paris: Anthropos, 1994.

Dauben, Joseph W. *Georg Cantor: His Mathematics and Philosophy of the Infinite*. Princeton, N. J.: Princeton University Press, 1990.

Dawson, John W., Jr. *Logical Dilemmas: The Life and Work of Kurt Gödel*. Natick, MA: A. K. Peters, 1997.

Drake, F. *Set Theory*. Amsterdam: North Holland, 1974.

Epstein, Perle. *Kabbalah: The Way of the Jewish Mystic*. New York: Barnes & Noble, 1998.

Field, J. V. *The Invention of Infinity: Mathematics and Art in the Renaissance*. NY: Oxford University Press, 1997.

Fraenkel, A., Y. Bar-Hillel, and A. Levy. *Foundations of Set Theory*. Amsterdam: North Holland, 1973.

Friedman, Avner. *Foundations of Modern Analysis*. New York: Dover, 1987.

Gödel, Kurt. *On Formally Undecidable Propositions of Principia Mathematica and Related Systems*. New York: Dover, 1962.

Gödel, Kurt. *Collected Works*. Vol. I. Feferman, Solomon, et al., eds. New York: Oxford University Press, 1986.

Gödel, Kurt. *Collected Works*. Vol. II. Feferman, Solomon, et al., eds. New York: Oxford University Press, 1990.

Grattan-Guinness, Ivor. *The Norton History of the Mathematical Sciences*. New York: Norton, 1998.

Gut, Emmy. *Productive and Unproductive Depression*. New York: Basic Books, 1989.

Hajnal, Andras, and Peter Hamburger. *Set Theory*. Cambridge, U.K.: Cambridge University Press, 1999.

Hallett, Michael. *Cantorian Set Theory and Limitation of Size*. NY: Oxford University Press, 1984.

Halmos, Paul R. *Naïve Set Theory*. New York: Van Nostrand, 1965.

van Heijenoort, J. ed. *From Frege to Gödel*. Cambridge, MA: Harvard University Press, 1967.

Hobson, Ernest Willaim. *Squaring the Circle*. Cambridge, U.K.: Cambridge University Press, 1913.

Hrbacek, K., and Thomas Jech. *Introduction to Set Theory*. New York: Marcel Dekker, 1978.

Kamke, E. *Theory of Sets*. New York: Dover, 1950.

Kanamori, Akihiro. *The Higher Infinite*. Berlin: Springer-Verlag, 1997.

Kelley, John. *General Topology*. Princeton, NJ: Van Nostrand, 1955.

Kline, M. *Mathematical Thought from Ancient to Modern Times*. New York: Oxford University Press, 1972.

Kunen, K. *Set Theory: An Introduction to Independence Proofs*. Amsterdam: North Holland, 1980.

Kuratowski, K. *Introduction to Set Theory and Topology*. Reading, MA: Addison-Wessley, 1962.

Lavine, Shaughan. *Understanding the Infinite*. Cambridge, MA: Harvard University Press, 1994.

Levy, A. *Basic Set Theory*. Berlin: Springer-Verlag, 1979.

Matt, Daniel C., Trans. *Zohar: The Book of Enlightenment*. Mahwah, N.J.: Paulist Press, 1983.

McLeish, John. *The Story of Numbers: How Mathematics Has Shaped Civilization*. New York: Fawcett, 1991.

Moore, Gregory H. *Zermelo's Axiom of Choice: Its Origins, Development, and Influence*. New York: Springer-Verlag, 1982.

Von Neumann, John. *Collected Works*. A.H. Taub, ed., Vol. 1. Oxford: Pergamon, 1961.

Phillips, Esther. *An Introduction to Analysis and Integration Theory*. New York: Dover, 1984.

Quine, W. V. O. *Set Theory and Its Logic*. Cambridge, MA: Harvard University Press, 1963.

Ramsey, Frank P. *The Foundations of Mathematics and Other Logical Essays*. R. Braithwaite, ed., London: Kegan Paul, 1931.

Rucker, Rudy. *Infinity and the Mind*. New York: Penguin, 1997.

Russell, Bertrand. *Autobiography, 1914-1944*. Boston: Little, Brown, 1968.

Shelah, Saharon. *Proper Forcing*. Berlin: Springer-Verlag, 1982.

Shelah, Saharon. *Cardinal Arithmetic*. New York: Oxford University Press, 1994.

Shoenfield, J. R. *Mathematical Logic*. Reading, MA: Addison-Wesley, 1967.

Schoenflies, A. *Entwickelung der Mengenlehre*. Leipzig: B. G. Teubner, 1913.

Sierpinski, Waclaw. *Hypothèse du Continu*. NY: Chelsea, 1956.

Steen, L. A., and J. A. Seebach. *Counterexamples in Topology*. New York: Dover, 1978.

Suppes, Patrick. *Axiomatic Set Theory*. New York: Van Nostrand, 1965.

Tarski, Alfred. *Logic, Semantics, Metamathematics: Papers from 1923 to 1938*. Oxford: Clarendon Press, 1956.

Wang, Hao. *A Logical Journey: From Gödel to Philosophy*. Cambridge, MA: MIT Press, 1996.

Wells, David. *The Penguin Dictionary of Curious and Interesting Numbers*. New York: Penguin, 1987.

Whitehead, Alfred North and Bertrand Russell. *Principia Mathematica*. Cambridge, U.K.: Cambridge University Press, Vol. 1, 1910; Vol. 2, 1912; Vol. 3, 1913.

Notes

1. See Carl Boyer, *A History of Mathematics,* New York: Wiley, 1968, p. 58.
2. Suppose that there are two integers, a and b, whose ratio is equal to the square root of two. Then $a^2= 2b^2$. Assume, without loss of generality, that the two integers are in lowest terms (they have no common factor, which could be canceled out). If a is odd, we have an immediate contradiction, because $2b^2$ is even. If a is even, it is equal to $2c$, for some number c. So we have $a^2=(2c)^2=4c^2$, which by the assumption must equal $2b^2$, thus b is even and hence a and b have the common factor 2, which is again a contradiction of our assumption.
3. There is a neat trick that proves that every number that has a repeating decimal sequence—no matter how long it is before it repeats itself—is a rational number. We'll demonstrate it with an example. Let's take the number 0.123123123123. . . . Call this number X. Now, $1000X=123.123123123123$; and $1000X-X=123.000$ (We got rid of the repeating decimals). Thus: $999X=123$, and therefore $X=123/999$, a ratio of two integers and hence a rational number.
4. See *Zohar: The Book of Enlightenment,* translated by D. C. Matt, Mahwan, N.J.: Paulist Press, 1983, p. 33.
5. Ibid., p. 49.
6. See R. Rucker, *Infinity and the Mind,* Princeton University Press, 1995, pp. 189-219.

7. Tan x stands for the trigonometric function tangent x, which for an acute angle x is the ratio of the side opposite the angle to the side adjacent to it, when considered part of a right triangle.
8. Even this assertion, among many others about Cantor's life, has been contested. The British historian of mathematics, Ivor Grattan-Guinness, claims in his article, "Towards a Biography of Georg Cantor," *Annals of Science*, Vol. 27, No.4, 1971, that Cantor was definitely not Jewish.
9. See E. T. Bell. *Men of Mathematics*. New York: Simon & Schuster, 1937.
10. Reprinted in Dauben, op. cit., pp. 275-6.
11. Dauben, op. cit., p. 277.
12. Paul R. Halmos. *Naïve Set Theory*. New York: D. Van Nostrand, 1960, p. v.
13. In Cantor's proof, the digit for each number is changed by a slightly different way than by adding one to it, but the principle is the same. For mathematically-minded readers, here is another proof that the numbers on the real line are uncountable. For simplicity, again we use just the interval from zero to one. Suppose that there is a way to enumerate all the numbers from zero to one on a list. Suppose the list is $a_1, a_2, a_3, \ldots$, and so on (an infinite list). Now, divide the interval 0 to 1 into three thirds: the closed interval from 0 to 1/3; the closed interval from 1/3 to 2/3; and the closed interval from 2/3 to 1. Choose one of these closed intervals that does not contain the first term in the sequence, a_1. Now divide this chosen interval again into thirds. Suppose the interval is 0 to 1/3. Now divide it into 0 to 1/9; 1/9 to 2/9; and 2/9 to 1/3. Choose one of these new intervals that does not contain the second number in the enumeration, a_2. Then continue in this way infinitely onward. Now, by a mathematical property, there is a point of intersection of any infinite collection of closed intervals. Suppose that point is called c. We know, by construction of the nested closed intervals, that this number is not one of the numbers in the sequence $a_1, a_2, a_3, \ldots$

14. Reprinted in Dauben (1979, p. 54).
15. Henri Lebesgue (1875-1941) was a French mathematician known for developing the theory of measure.
16. Dauben, p. 1.
17. Hallett, Michael. *Cantorian Set Theory and Limitation of Size.* NY: Oxford University Press, 1984, p. 13.
18. Letter from Cantor to Gösta Mittag-Leffler, Dec. 30, 1883, reprinted in Dauben (1979, p. 134).
19. David Wells, *The Penguin Dictionary of Curious and Interesting Numbers,* London, U.K.: Penguin, 1987, p. 205.
20. St. Augustine. *City of God.* New York: Penguin, 1972, pp. 496-7.
21. The discontinuous functions are the ones giving this set its higher order of infinity. A popular book about mathematics describes this order of infinity as: "all the curves you can draw on the back of a stamp." This statement is incorrect not because of the small size of the stamp (we already know that size as we know it doesn't affect infinity), but rather because curves alone—continuous drawings on the stamp—will not do. Continuous functions have the order of infinity of the real numbers. The discontinuous functions are of a higher order.
22. These numbers, both finite and transfinite, are called ordinal numbers. In what follows we will discuss the more interesting infinite (and finite) numbers called cardinal numbers, which comprise the most important element of Cantor's work and modern mathematics.
23. I. Grattan-Guinness, "Towards a Biography of Georg Cantor," *Annals of Science,* 27, No.4, 1971, p. 351. The word *not* is italicized in the paper.
24. Nathalie Charraud. *Infini et Inconscient: Essai sur Georg Cantor.* Paris: Anthropos, 1994, p. 176.
25. A. Schoenflies. *Entwickelung der Mengenlehre.* Leipzig: B. G. Teubner, 1913.
26. I. Grattan-Guinness, "Towards a Biography of Georg Cantor," *Annals of Science,* 27, 1971, p. 372.

27. *The Autobiography of Bertrand Russell.* London, 1967-69 (published in the U.S. by Bantam, 1968), vol. I, p. 217; Cantor's letters are reproduced on pp. 218-220.
28. Dauben (1990), p.239.
29. This paradox and others are discussed in Patrick Suppes. *Axiomatic Set Theory.* New York: Van Nostrand, 1965, p. 9.
30. Adapted from Willard V. O. Quine. *Set Theory and Its Logic.* Cambridge, MA: Harvard University Press, 1963, p. 254.
31. Reprinted in Patrick Suppes. *Axiomatic Set Theory.* New York: D. Van Nostrand, 1960, p. 5 n.
32. The details about Gödel's life and work are from John W. Dawson's *Logical Dilemmas: The Life and Work of Kurt Godel.* Wellesley, MA: A. K. Peters, 1997; Hao Wang's *A Logical Journey: From Godel to Philosophy.* Cambridge, MA: M.I.T. Press, 1996; and S. Feferman, et al., eds., The Collected Works of Kurt Godel, Vols. I-III, New York: Oxford University Press, 1990.
33. Dawson, 1997, p. 16.
34. Quoted in Dauben, 1990, p. 290.
35. A functional $F(x)$ is linear if $F(x+y)=F(x)+F(y)$, and homogeneous if $F(ax)=aF(x)$.
36. In actuality Gödel proved a set of theorems; one is described here, for simplicity.
37. Hallett, Michael. *Cantorian Set Theory and Limitation of Size.* New York: Oxford University Press, 1984, pp. 7-11.
38. It should be noted that Cantor and Gödel were not the only mathematicians working in the field of the foundations of mathematics to suffer from mental illness—depression and a persecution complex. Ernst Zermelo, the developer of the system of axioms and the father of the axiom of choice, also suffered from at least one nervous breakdown. The mathematician Emil L. Post, who had anticipated some of Gödel's results about infinity but missed proving them himself, suffered from the same illness. It is interesting to contemplate the reasons for this striking anomaly.

39. The Nazis were not the only ones to make this mistake. In his *Autobiography,* Bertrand Russell described Gödel as a Viennese Jew—one of a number of mistakes the English philosopher made in his book.
40. John W. Dawson. *Logical Dilemmas: The Life and Work of Kurt Gödel.* Wellesley, MA: A. K. Peters, 1997, p. 109.
41. Dawson, op. cit.
42. Dauben, op. cit., pp. 269-70
43. See a translation of Cantor's 1883 paper in *Mathematische Annalen* in Michael Hallett. *Cantorian Set Theory and Limitation of Size.* NY: Oxford University Press, 1984, p. 17.

Two pages of Cantor's notes leading to the concept of actual infinity.

Befreundete Zahlen

Beweis d. Cartesischen Formel

Statt 2A kann man, da A eine Potenz v. 2 ist, setzen 2^{m+1} so ist $2AR = 2^{m+1}R$
$2APQ = 2^{m+1}PQ$.

1, Die Teilersumme v. $2AR$ [illegible] da 2^{m+1} [illegible] einfache Teiler als 2, [illegible] R eine Primzahl ist, [illegible] $2AR$ [illegible] Teiler als 2 [illegible] R [illegible]

1	1
1	2
1	2^2
1	2^3
	.....
1	2^{m+1}
R	1
R	2
R	2^2
R	2^{m+1}

Die Summe aller Teiler ist [illegible]

$$1+2+2^2\ldots 2^{m+1} + R(1+2+2^2\ldots 2^m)$$
$$= 2^{m+2}-1+R(2^{m+1}-1)$$
$$= 2^{m+2}-1+(18A^2-1)(2^{m+1}-1)$$
$$= 2^{m+2}-1+(18.2^{2m}-1)(2^{m+1}-1)$$
$$= 2^{m+2}-1+18.2^{3m+1}-18.2^{2m}-2^{m+1}+1$$
$$= 2^{m+2}+18.2^{3m+1}-18.2^{2m}-2^{m+1}$$
$$= 2^{m+1}(2+18.2^{2m}-9.2^m-1)$$
$$= 2^{m+1}(1+18.2^{2m}-9.2^m)$$
$$= 2^{m+1}(1+18A^2-9A) = 2^{m+1}(3A-1)(6A-1)$$
$$= 2A.P.Q$$

[illegible] $A = 2^m$ $(2^l+1)A-1 = P$. $2^l(2^l+1)A-1 = Q$. $2^l(2^l+1)^2A^2-1 = R$. [illegible] $2^{m+1}R$ [illegible] $2^{m+1}PQ$ [illegible]

also ist d. Summe d. Teiler v. $2AR = 2A.P.Q$. Ebenso ist

Index